Topology Design of Robot Mechanisms

Ting-Li Yang · Anxin Liu
Huiping Shen · Lubin Hang
Yufeng Luo · Qiong Jin

Topology Design of Robot Mechanisms

Springer

Ting-Li Yang
Changzhou University
Changzhou
China

Lubin Hang
Shanghai University of Engineering Science
Shanghai
China

Anxin Liu
Nanhang Jincheng College
Nanjing
China

Yufeng Luo
East China Jiaotong University
Nanchang
China

Huiping Shen
Changzhou University
Changzhou
China

Qiong Jin
The Boeing Company
California
USA

ISBN 978-981-10-5531-7 ISBN 978-981-10-5532-4 (eBook)
https://doi.org/10.1007/978-981-10-5532-4

Library of Congress Control Number: 2017951419

Printed on acid-free paper

This Springer imprint is published by Springer Nature
The registered company is Springer Nature Singapore Pte Ltd.
The registered company address is: 152 Beach Road, #21-01/04 Gateway East, Singapore 189721, Singapore

Preface

Robot mechanisms, especially parallel robot mechanisms, have raised great interest among many mechanism researchers during the last two decades. Compared with traditional serial mechanisms, parallel robot mechanisms offer many advantages, such as high accuracy, rigidity, load-to-weight ratio, etc. Early research of parallel robot mechanisms was focused mainly on the six-DOF Gough–Stewart platform. In the last decade, parallel mechanisms with fewer DOFs attracted the attention of researchers and gradually became popular in many fields.

There are two principal goals in researching mechanism topology: One is to provide systematic theories and methods, which eventually guide the topology design of robot mechanisms. The other is to establish new composition principles of mechanisms, based upon which a unified modeling of topology, kinematics, and dynamics of mechanisms can be achieved.

This book introduces the creative research achievements of the authors on mechanism topology over the past more than thirty years. The essential points can be concentrated as follows:

(1) **Three fundamental concepts**

   (a) Geometrical Constraint Type of axes (see Chap. 2).
   The subject concept describes the geometrical constraints among axes of kinematic pairs. The geometrical constraint type of axes, together with the kinematic pair type, and the connection relations among links are three basic elements of topological structures. It should be emphasized that the symbolic expression of a topological structure constructed in this manner is independent of the motion positions of a parallel mechanism (excluding singular positions).

   (b) POC Set (see Chap. 3)
   The POC set, derived from the unit vector set of the velocity of a link, the POC set describes the position and orientation characteristics of the relative motions between any two links of a mechanism. It depends only on the topological structure of the mechanism and is independent of the motion positions of the mechanism (excluding singular positions).

(c) SOC Unit (see Chap. 2)
The SOC unit is used as the basic structure unit to develop four basic equations of mechanism topology (see Chaps. 4–6).

(2) **Mechanism composition principle based on SOC units**
The mechanism composition principle based on SOC units is proposed to establish the systematic theory for unified modeling of topology, kinematics, and dynamics of mechanisms (see Chap. 7).

(3) **Four basic equations**
The equations listed in the following Paragraphs (a) through (c) reveal the mapping relationship among the topological structure, the DOF and the POC set of a mechanism. The equation addressed in the following Paragraph (d) provides the theoretical cornerstone for building up a systematic theory and method for topology, kinematics, and dynamics of mechanisms based on the SOC-unit modeling.

(a) POC Equation of Serial Mechanisms (see Chap. 4)
This equation is used for topological structure analysis and synthesis of serial mechanisms. Eight linear symbolic operation rules and two nonlinear criteria are created to support this equation.
(b) POC Equation of Parallel Mechanisms (see Chap. 5)
This equation is used for topological structure analysis and topology design of parallel mechanisms. Twelve linear symbolic operation rules and two nonlinear criteria are provided for this equation.
(c) The General DOF Formula (see Chap. 6)
This is the formula to calculate the DOF of parallel mechanisms and multi-loop spatial mechanisms.
(d) The Coupling Degree Formula for AKC (see Chap. 7)
This is the formula to calculate the coupling degree of the Assur kinematic chain (AKC). The coupling degree represents the complexity of kinematic and dynamic analyses of the AKC.

The above four equations (formulas) depend only on the topological structures of a mechanism and are independent of motion positions of the mechanism (excluding singular positions). It is not necessary to establish a fixed coordinate system when using these four equations.

(4) **One systematic method for topology design of parallel mechanisms** (see Chaps. 8–10)
Based on the three fundamental concepts, the mechanism composition principle based on SOC units, and the four basic equations mentioned in Paragraphs (1), (2), and (3). The systematic topology design of parallel mechanisms, developed from the essential points (1), (2), and (3) mentioned above, possesses the following characteristics:

(a) The design has two stages: Stage 1 synthesizes structures obtaining many kinds of parallel mechanisms; Stage 2 analyzes the performances, classifies types, and subsequently optimizes the parallel mechanisms synthesized from Stage 1.

(b) The design operation is independent of the motion positions of a mechanism (excluding singular positions) and the fixed coordinate system (i.e. it is not necessary to establish the fixed coordinate system); therefore it is a geometrical method. This ensures a full-cycle DOF of the parallel mechanism and the generality of geometric conditions of assembling branches between the two platforms.
(c) Each individual design step follows an explicit formula or the guidelines of design criteria, making the operation simple and feasible.

It may be beneficial to outline the following chronological milestones of our researches.

A. 1983–1986: A general method for structural synthesis of overconstrained spatial single-loop mechanisms was provided, based on which new structure types were synthesized (Ref. [2] in Chap. 1, and Ref. [9] in Chap. 2). The geometrical constraint type of axes (the type of geometrical constraint on axes of adjacent kinematic pairs, which was originally called the dimensional constraint type) was introduced as the third key element to describe the topological structure of a mechanism. The other two key elements are: the kinematic pair and the connection relation between links. The mapping relationship between the number of independent displacement equations and the topological structure of overconstrained spatial single-loop mechanisms was revealed based on these three key elements.

B. 1985–1996: The composition principle based on the SOC units for planar mechanisms was proposed. An SOC-based structure synthesis method for planar mechanisms was presented, and the corresponding structural types were obtained (Ref. [59] in Chap. 1). The key points are as follows: (a) the SOC-based structure unit, (b) the constraint degree formula for a planar SOC, and (c) the coupling degree formula for the planar AKC. These three points were later used to form the systematic theory and method for unified modeling of topology, kinematics, and dynamics of planar mechanisms (Ref. [59, 61, 62, 65, 69] in Chap. 1). This theory and method was described in detail in the first author's monograph, which was written in Chinese, "Basic Theory of Planar Mechanisms: Structure, Kinematics and Dynamics," China machine press, Beijing, 1996 (Ref. [61] in Chap. 1). It is well-known and frequently cited by Chinese scholars.

C. 1995–2005: An SOC-based structure synthesis method for parallel mechanisms was given, and structural types with DOF=2–6 were derived (Ref. [3, 27, 28–30, 31, 32, 33–37] in Chap. 1). The innovations of this phase are as follows:

(1) It was discovered that the topological structures constructed by the three basic elements mentioned in the Milestone (A) is motion process invariant (excluding singular positions).
(2) The POC set, originally named "POC Matrix" and derived from the unitary vector set of the link velocity, depicts the relative movement characteristics of any two links. It is independent of the motion positions (excluding singular positions) and the fixed coordinate system.

(3) The POC equation for a serial mechanism and its operation rules are independent of the motion positions and the fixed coordinate system.
(4) The POC equation for a parallel mechanism and its operation rules are independent of the motion positions and the fixed coordinate system.

Published in 2004, our Chinese renowned treatise, "Theory of Topological Structure for Robot Mechanisms" China machine press, Beijing (Ref. [32] in Chap. 1), describes the systematic theory and application of the structural synthesis of parallel mechanisms. This is the first academic book in the world on structural synthesis of parallel mechanisms.

D. 2006–2016: The innovations during this period are as follows:

(1) A general DOF formula for parallel mechanisms and arbitrary spatial multi-loop mechanisms was gradually established (Refs. [40-42] in Chap. 1). The creative points are: (a) a full-cycle DOF obtained, (b) a criterion to determine inactive kinematic pairs, (c) a criterion to determine driving pairs, and (d) a simplified method, based on the topological equivalent principle, to calculate the DOF of a mechanism.
(2) Using this general DOF formula for spatial mechanisms, the composition principle based on SOC units was able to be extended to general spatial mechanisms. The constraint degree formula for a spatial SOC and the coupling degree formula for the spatial AKC are presented (Ref. [43] in Chap. 1).
(3) A general method for topology design of PMs is proposed. The design process includes two stages: (a) Structure synthesis. Many structure types are obtained. (b) Performance analysis, classification and optimization of PM structure types (refer to Chaps. 9–10).
(4) It provides a theoretical basis for establishment of a unified SOC-based method for structure synthesis and kinematic (dynamic) analysis of general spatial mechanisms (Ref. [43] in Chap. 1).

E. 2013: The paper "On the Correctness and Strictness of the POC Equation for Topological Structure Design of Robot Mechanisms" was published in JMR (Ref. [39, 38] in Chap. 1) to answer the questions or doubts of some researchers on our theory: (a) Is this theory correct or rigorous? and (b) Can this theory be used to solve any topology design problems which cannot be solved by other existing methods?

In conclusion, the five milestones summarized above are foundations of the systematic theory and methodology demonstrated in this book.

Provided below is a brief of the idea of my research on the theory mentioned above.

In 1959, when I was a junior at Qinghua University (Beijing) studying the course "Theory of Machines," I had quite a few doubts about the DOF formula for spatial mechanisms. Meanwhile, I was deeply interested in the Assur structure theory. In this respect, my long-term research on the subject originated from the doubts and curiosity of mine at that time. However, I was not able to concentrate my research on the subject until the Ten-year Catastrophe (1966–1976) which is also known as the Cultural Revolution was over. If my research is officially counted from my presenting the first research paper on the subject at the sixth IFToMM world congress on the theory of machines and mechanisms held in New Deli in 1983, it has been thirty-five years—never a short term in anyone's life!

Introduction of the three new concepts based on non-logical thinking enables the theory to clearly depict its physical significance, and the four basic formulas derived from logical reasoning give the theory a precise mathematical structure. As a result, the theory well demonstrates the beauty of simplicity—a perfect reflection of the Daoist philosophy "Greatest truths are always the simplest."

As the beauty of simplicity in theories, the trueness of emotion, and the rich colorfulness of the Mother Nature are closely linked, appreciation and understanding of and resonance to such beauties further stimulate new instinct and inspiration.

In addition, adoration of the spirit of independence and freedom of thought, extensive reading with reflection, emotion and sense in making friends, and thinking of the light or darkness in the social development enable people to have a broader vision and promote in-depth thinking. The passion and aesthetic feeling generated from life have become a pillar of spirit in my long-lasting concentrated academic study.

As the lead author, I am deeply grateful for the invaluable encouragement and help from Prof. Shi-Xian Bai (Beijing Polytechnic University), Prof. Wei-Qing Cao (Xi'an University of Technology), Prof. Qi-Xian Zhang (Beijing university of Aeronautics and Astronautics), and Prof. Xi-Kai Huang (Southeast University, Nanjing).

I would also like to express my sincere thanks to Prof. Ming Zhang, Prof. Meiyu Lin, Senior Engr. Zhen Xu, Engr. Dongjin Sun, Engr. Fanghua Yao, and other friends. The broad communication with these people, which often shines lights on me resulting in new trains of thoughts, has been a spiritual pleasure indeed in my entire research career.

Great thanks also go to Dr. Yufeng Luo, Dr. Anxin Liu, Dr. Huiping Shen, Dr. Qiong Jin, Dr. Lubin Hang, Dr. Xianwen Kong, Dr. Zhixin Shi, Prof. Jinkui Chu, Prof. Chuanhe Liu, Prof. Zhiyou Feng, MSc. Huiliang Li, MSc. Jianqing Zhang, MSc. Baoqian Shi, MSc. Songqin Shan and many other scholars for their continuous contributions and assistances to me in the past three decades.

The authors would like to acknowledge the financial support of the National Nature Science Foundation of China.

Finally, my special gratitude goes out to my wife Jingying Shao and other family members. Without their full support during the past more than forty years, this book would not have been possible.

Nanjing, China
March 28, 2017

Ting-Li Yang

Contents

Abbreviations

AKC	Assur kinematic chain
$AKC_i[v_i]$	the ith AKC with v_i independent loops
C	Cylindrical pair
dim (M)	Dimension of POC set, i.e. the number of independent elements in the POC set
dim $(M(r))$	Number of independent rotational elements
dim $(M(t))$	Number of independent translational elements
DOF	Degree of freedom
D_{red}	Redundancy of the mechanism
e_R	Unit vector in direction of R-axis
e_P	Unit vector in direction of P-axis
e_ρ	Unit vector of radius vector
F	Degree of freedom
f_i	DOF of the *i*th joint
H	Helical pair
HSOC	Hybrid SOC, i.e. complex SOC with loop
KC	Kinematic chain
$KC[F, v]$	Kinematic chain with DOF=F and v independent loops
$M(\dot{M})$	POC set (VC set)
$M_b(\dot{M}_b)$	POC set (VC set) of branch
$M_H(\dot{M}_H)$	POC set (VC set) of H pair
$M_P(\dot{M}_P)$	POC set (VC set) of P pair
$M_{Pa}(\dot{M}_{Pa})$	POC set (VC set) of PM
$M_R(\dot{M}_R)$	POC set (VC set) of R pair
$M_S(\dot{M}_S)$	POC set (VC set) of SOC
$M_{S(L)}$	POC set of $SOC_{(SLC)}$
$M_{\mathrm{sub}-SOC_j}$	POC set of the *j*th sub-SOC
m	Number of joints of mechanism
m_j	Number of joints of the *j*th branch
N_{ov}	Number of overconstraints

n	Number of links
P	Prismatic pair
PM	Parallel mechanism or parallel manipulator
POC	Position and orientation characteristics
P^*	Translation of a P pair or derivative translation of an R (or H) pair
$P^{(4R)}$	One translation generated by a parallelogram formed by four R pairs
$P^{(4S)}$	One translation generated by a parallelogram formed by four S pairs
R	Revolute pair
$r^1(\parallel R)$,$r(\parallel R)$	A rotation in a direction parallel to axis of R pair
$r^1(\parallel H)$,$r(\parallel H)$	A rotation in a direction parallel to axis of H pair
$r^2(\parallel \diamondsuit(R, R^*))$	Two rotations in directions within a plane parallel to axis of R pair and R^*pair
r^3	Three rotations
$\{r^i(dir)\}$	Non-independent element put into a brace
S	Spherical pair
SLC	Single-loop chain
SOC	Single-Open-Chain, i.e. serial mechanism
$\mathrm{SOC}_{(\mathrm{SLC})}$	SOC corresponding to SLC (Sect. 4.6)
$\mathrm{SLC}_{(\mathrm{SOC})}$	SLC corresponding to SOC (Sect. 4.6)
$t^1(\parallel P)$, $t(\parallel P)$	One translation in a direction parallel to axis of P pair
$t^1(\perp R_1)$, $t(\perp R)$	One translation in a direction within a plane perpendicular to axis of R pair
$t^1(\parallel^{\perp} (ad))$	One translation may be parallel to line (ad) or may be perpendicular to line (ad)
$t^2(\perp R)$	Two translations in directions within a plane perpendicular to axis of R pair
$t^2(\perp \rho)$	Two translations in directions within a plane perpendicular to radius vectorρ
$t^2(\parallel \diamondsuit(P, P^*))$	Two translations in directions within a plane parallel to P and P^*
t^3	Three translations
$\{t^i(dir)\}$	Non-independent element put into a brace
U	Universal pair
VC	Velocity characteristics
ξ	Number of independent displacement equations of KC
ξ_L	Number of independent displacement equations of SLC
ξ_{L_j}	Number of independent displacement equations of the jth loop
Δ_j	Constraint degree of the jth SOC
κ	Coupling degree of AKC
v	Number of independent loops
$R \parallel R \parallel \cdots \parallel R$	Axes of several adjacent R (or H) pairs are parallel (Fig. 2.2a)
$R \nparallel R$	Axes of two R pairs are not parallel
$R \vert H$	Axes of two adjacent pairs are coincident (Fig. 2.2b)

$\overbrace{RR\cdots R}$	Axes of several adjacent R pairs intersect at one point (Fig. 2.2c)
$R\perp P$	Axes of two adjacent pairs are perpendicular (Fig. 2.2d)
$R\not\perp P$	Axes of two pairs are not perpendicular
$\Diamond(P, P, \cdots, P)$	Axes of several P pairs are parallel to one plane (Fig. 2.2e)
$R - R$	Axes of kinematic pairs are allocated arbitrarily in the space (Fig. 2.2f)
$R_1(\perp P_2) \parallel R_3$	$R_1 \perp P_2$ and $R_1 \parallel R_3$
$R \parallel^{\perp} P^{(4R)}$	Axes of R pair and $P^{(4R)}$ may be parallel or perpendicular
$R \parallel^{\perp} P^{(4S)}$	Axes of R pair and $P^{(4S)}$ may be parallel or perpendicular
$\Diamond(R_i, R_j)$	A plane parallel to axes of R_i pair and R_j pair
$\Diamond(4R)$	Four R pairs form a parallelogram
$\Diamond(4S)$	Four S pairs form a parallelogram all the time

Chapter 1
Introduction

1.1 Robot Mechanism Topology

1.1.1 Topological Structure Design of Mechanisms

Topological structure design of a mechanism is to determine the three basic topological elements and the schematic diagram of the mechanism. The three essential topological elements of a mechanism are:

(1) Types and number of kinematic pairs [1] (see Sect. 2.2.1 for detail).

(2) Connection relations among links and kinematic pairs [1] (see Sect. 2.2.2 for detail).

(3) Types of geometrical constraint between kinematic pair axes attached by links [2–4] (see Sect. 2.2.3 for detail).

Traditional topological structure of mechanisms considers only the first two basic elements [1]. "Types of geometrical constraint between kinematic pair axes' was introduced into mechanism topology as the third basic element in 1983 (originally called as "dimensional constraint type") [2, 3]. It is necessary to introduce the third basic element. Otherwise it will be hard to describe the different topological structures of the two mechanisms shown respectively in Fig. 2.4a, b.

In this book, the systematic theory and method for topological structure design of mechanisms is called mechanism topology and the topological structure design of mechanisms is abbreviated as mechanism topology design.

1.1.2 Basic Tasks of Mechanism Topology

Fundamental tasks of mechanism topology are: (1) to provide a systematic theory and method for mechanism topology design (i.e. invention of new mechanisms) and

T.-L. Yang et al., *Topology Design of Robot Mechanisms*,
https://doi.org/10.1007/978-981-10-5532-4_1

(2) to set up new mechanism composition principle and, based on this principle, establish a systematic theory and method for unified modeling of mechanism topology, kinematics and dynamics.

1.1.3 Basic Mode of Mechanism Topology Design

Generally, the mechanism topology design follows a 3-stage mode:

⟪Task space⟫	→	⟪Solution space⟫	→	⟪Solution optimization⟫
(Basic functions)		(Set of structure types)		(Structure type optimization)

- Task space: to determine the design objectives based on the design specification, including functional requirements and some performance requirements.
- Solution space: to obtain many structure types which meet the basic functional requirements using the structure synthesis method and thus to provide a relatively larger solution space (a set of structure types) for structure type optimization.
- Solution optimization: to carry out performance analysis, comparison and classification to the structure types in the solution space based on functional requirements, some performance requirements and other specific requirements and to provide some useful information for structure type optimization.

Therefore, mechanism topology design shall include two aspects: structure synthesis (to obtain many structure types) and performance analysis (to classify the obtained structure types according to their performance).

1.1.4 Main Methods for Mechanism Topology Design

In the decade, topology design of parallel robot mechanism became the research focus of many researchers. There formed the following four categories of topology design methods:

(1) Method based on screw theory [5–12]

Main features of this method as follows:

(a) Six components of plucker coordinates are used to describe relative motion characteristics between two links of a mechanism. They are dependent on motion position of the mechanism and the fixed coordinate system.
(b) Linear operation of the screw system is relatively simple.

   (c) Complex "intersection" operation among motion screw systems is converted to simple "union" operation among constraint screw systems by using reciprocal product of screws.
   (d) For the obtained parallel mechanisms, it is necessary to carry out the full-cycle DOF verification since this method depends on motion position of a mechanism.

(2) Method based on subgroup/submanifold [12–22]

   Main features of this method as follows:

   (a) 12 kinds of displacement subgroups are used for describing the relative motion characteristics of a mechanism fulfilling the algebra structures of Lie group.
   (b) "Union" and "intersection" operation rules of displacement subgroups are provided using synthetic arguments or tables of compositions of subgroups [13, 15, 22].
   (c) The full-cycle DOF mechanisms are obtained.
   (d) For mechanisms not fulfilling the algebra structures of Lie group, such as having two dimensional rotations, or five dimensional motion etc [12], the displacement submanifold based method can be used for structure analysis and synthesis, which depends on motion position of the mechanism [21]. Therefore, it is necessary to carry out the full-cycle DOF verification for mechanisms obtained by this method.

(3) Method based on theory of linear transformations and evolutionary morphology [22–28].

   Main features of this method as follows:

   (a) Velocity space is used to describe relative motion characteristics between two links of a mechanism. It is dependent on motion position of the mechanism and the fixed coordinate system.
   (b) The design objectives include DOF of branch, DOF of PM, connectivity of branch between the moving platform and the fixed platform, connectivity of PM between the moving platform and the fixed platform, number of over constraints and redundancy of the mechanism.
   (c) Morphological operators (the combination, the mutation, the migration and the selection) and evolutionary rules are used to determine type, number and order of kinematic pairs and allocation of pair axes.
   (d) For the obtained parallel mechanisms, it is necessary to carry out the full-cycle DOF verification since this method depends on motion position of a mechanism.

(4) Method based on POC equations [4, 29–47].

   Main features of this method as follows:

   (a) Six elements of POC set are used to describe relative motion characteristics between two links of a mechanism. They are independent of motion position (excluding singular position) of the mechanism and the fixed coordinate system.

(b) The topology design process is divided into two stages: the structure synthesis is conducted and many structure types are obtained during the first stage and performance analysis and classification are carried out for those obtained structure types during the second stage.

(c) Theoretical bases of this method are the POC equation for serial mechanisms, the POC equation for parallel mechanisms and the general DOF formula. These equations (formula) are independent of motion position of a mechanism and the fixed coordinate system (i.e. it is not necessary to establish the fixed coordinate system).

(d) Since this method is independent of motion position, the full-cycle DOF mechanisms are obtained. Since the method is independent of the fixed coordinate system, the geometry conditions of mechanism existence have generality.

(e) Since this method is independent of motion position of a mechanism and the fixed coordinate system, it could be called a geometrical method, which is totally different from the other three methods.

1.1.5 Mechanism Composition Principles and Mechanism Theory Systems

In the development history of mechanism theory, there formed 4 types of mechanism composition principles. Corresponding theory systems covering topology, kinematics and dynamics of mechanisms have been established based on these composition principles.

(1) Mechanism composition principle based on AKC (i.e. Assur kinematic chain). The corresponding systematic theory and method for topology, kinematics and dynamics of mechanisms based on AKC modeling are established [48–52].

(2) Mechanism composition principle based on link-pair unit. The corresponding systematic theory and method for topology, kinematics and dynamics of mechanisms based on link-pair unit modeling are established [50, 53–56].

(3) Mechanism composition principle based on loop unit. The corresponding systematic theory and method for topology, kinematics and dynamics of mechanisms based on loop unit modeling are established [50, 57–59].

(4) Mechanism composition principle based on SOC unit. The corresponding systematic theory and method for topology, kinematics and dynamics of mechanisms based on SOC unit modeling are established [44, 60–73].

1.2 Objectives and Originality of This Book

1.2.1 Objectives

Our original work on robot mechanism topology during the past 30 years is introduced in this book. The main objectives are:

(1) to provide an original systematic method for topology design of robot mechanisms. This topology design method covers two aspects—structure synthesis (to obtain many structure types) and performance analysis (to select the best structure type) and has been used in topology design of (2T-0R), (1T-1R), (0T-2R), (3T-0R), (0T-3R), (2T-1R), (1T-2R), (3T-1R), (2T-2R), (1T-3R), (3T-2R), (2T-3R) and (3T-3R) parallel mechanisms.

(2) to provide a systematic theory and method for topology, kinematics and dynamics of mechanisms based on SOC unit modeling.

Main contents of this book include 3 basic concepts for mechanism topology, 1 mechanism composition principle, 4 basic equations, 12 topological characteristics and a general method for PM topology design.

1.2.2 Basic Concepts

Three new basic concepts for mechanism topology are proposed.

(1) The geometric constraint type of axes.
Type of geometric constraint relation between kinematic pair axes (abbreviated as geometric constraint type of axes). [2, 3], together with the other two existing concepts (type of kinematic pair, connection relation between links) [1] are used to describe the mechanism topological structure. They are called three basic elements of mechanism topological structure. Topological structure described by the three key elements is independent of motion position of the mechanism (excluding singular positions) and the fixed coordinate system. So, it is called the topological structure invariance during motion process. Refer to Chap. 2 for detail.

(2) Position and orientation characteristics (POC) set.
The POC set is used for describing the relative motion characteristics between any two links of a mechanism. The POC set is defined based on the unit vector set of velocity component set of link, which depends only on topological structure of a mechanism and is independent of motion position (excluding singular positions) of the mechanism and the fixed coordinate system. Refer to Chap. 3 for detail [3, 4, 29–36].

(3) Single open chain (SOC) unit.
The SOC unit is used as the basic structure unit to establish the basic topology equations (refer to Chaps. 4, 5 and 6) and to develop a new mechanism composition principle (Chap. 7).

1.2.3 Mechanism Composition Principle

Mechanism composition principle based on SOC unit is proposed and is used to build up a systematic theory and method for topology, kinematics and dynamics of mechanisms (refer to Chap. 7 for detail).

1.2.4 Basic Equations

Based on the above three new concepts, four basic equations (formula) for mechanism topology are established.

(1) POC equation for serial mechanisms (Eq. 4.4) [3, 4, 33, 39, 40]

$$M_S = \bigcup_{i=1}^{m} M_{J_i} \tag{4.4}$$

There are 10 symbolic operation rules for Eq. (4.4) (eight linear rules and two nonlinear criteria). The equation and these operation rules are independent of motion position of the mechanism (excluding singular position) and the fixed coordinate system (refer to Chap. 4 for detail).

(2) POC equation for parallel mechanisms (Eq. 5.3) [3, 4, 33, 39, 40]

$$M_{Pa} = \bigcap_{j=1}^{(v+1)} M_{b_j} \tag{5.3}$$

There are 14 symbolic operation rules for Eq. (5.3) (12 linear rules and two nonlinear criteria). The equation and these operation rules are independent of motion position of the mechanism (excluding singular position) and the fixed coordinate system (refer to Chap. 5 for detail).

(3) General formula for DOF calculation (Eq. 6.8) [41–43].

$$\begin{cases} F = \sum\limits_{i=1}^{m} f_i - \sum\limits_{j=1}^{v} \xi_{L_j} \\ \xi_{L_j} = \dim\left\{ \left(\bigcap_{i=1}^{j} M_{b_i} \right) \cup M_{b_{(j+1)}} \right\} \end{cases} \tag{6.8}$$

The above three Eqs. (4.4), (5.3) and (6.8) reveal the mapping relations among topological structure, POC and DOF of a mechanism (Fig. 1.1). They are basic equations for mechanism topology analysis and synthesis (refer to Chaps. 6–10 for detail).

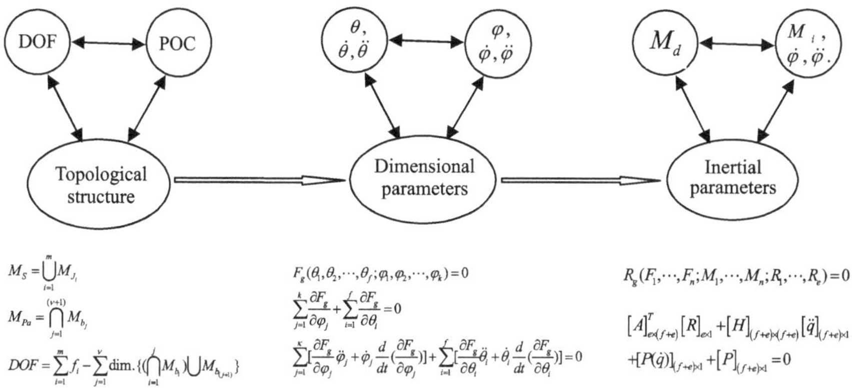

Fig. 1.1 Mapping relations among basic equations

(4) Formula for AKC's coupling degree calculation (Eq. 7.9) [3, 4, 33, 44, 46, 60, 61].

$$\kappa = \frac{1}{2}\min\left\{\sum_{j=1}^{v}|\Delta_j|\right\} \tag{7.9}$$

The coupling degree formula (Eq. 7.9) reveals the mapping relations among mechanism topological structure, SOC constraint degree and dimension of AKC's kinematic (dynamic) equations (i.e. number of unknowns). It provides a theoretical basis for building up a systematic theory and method for topology, kinematics and dynamics of mechanisms based on SOC unit modeling (refer to Chap. 7 for detail).

1.2.5 Topological Characteristics of Parallel Mechanisms

Based on the above four basic equations (formula), a set of topological characteristics of parallel mechanisms can be derived. These topological characteristics include POC set, DOF, number of independent displacement equations, number of over constraints, redundant DOF, inactive kinematic pairs, driving pairs allocation, type and number of AKCs, coupling degree of AKC, type of DOF, motion decoupling, etc. They have close relations with basic functions, kinematic (dynamic) performance and control performance of parallel mechanisms (Chap. 9) and can be used in performance analysis, classification and optimization of parallel structure types (Chap. 10).

1.2.6 Topology Design of Parallel Mechanisms

Based on the above four basic equations (formula) and those topological characteristics, a systematic method for topology design of parallel mechanisms is established. The design process covers two aspects: structure synthesis based on POC equations and performance analysis based on topological characteristics. The key steps include:

(1) Structure synthesis of simple branches (i.e. SOC branches) based on POC equation for serial mechanisms. Many different structure types of simple branches are obtained in this step (Chap. 8).

(2) Structure synthesis of complex branches based on topological equivalence principle. Many different structure types of complex branches are obtained in this step (Chap. 9).

(3) Arranging combination schemes of branches according to the types of branches and the number of branches in the PM to be designed (Chap. 9).

(4) Determining the geometric conditions for assembling branches between two platforms based on the POC equation for parallel mechanisms and the DOF formula. For a certain geometric condition, there may be several different schemes for assembling the branch between two platforms (Chap. 9).

(5) Performance analysis and classification of those obtained PM structure types based on topological characteristics of PM. Result of the analysis and classification may be used for structure type assessment and optimization (Chaps. 9 and 10).

Since the above four equations are independent of motion position of a mechanism and the fixed coordinate system, this topology design method is also independent of motion position of a mechanism (excluding singular positions) and the fixed coordinate system. So, the full-cycle DOF is obtained and geometry conditions of mechanism existence is general. Therefore, it is a a geometrical method, which is totally different from the other three methods.

Recently, geometric algebra (GA) is introduced to describe the POC of robot mechanisms [72]. The POC union is defined by the outer product operation of GA and the POC intersection is defined by the shuffle product operation of GA.

1.2.7 Application

Topology design of (3T-1R) PM is discussed in detail in Chap. 10. This topology design is divided into two stages. The first stage involves structure synthesis, in which many structure types of (3T-1R) PM are obtained. The second stage covers performance analysis and classification of (3T-1R) PMs based on topological characteristics.

The topology design method based on POC equations has been used in topology design of (2T-0R), (1T-1R), (0T-2R), (3T-0R), (0T-3R), (2T-1R), (1T-2R), (3T-1R), (2T-2R), (1T-3R), (3T-2R), (2T-3R), and (3T-3R) PMs. Many new PM structure types have been obtained [4, 28–37, 45, 46].

1.2.8 Mechanism Theory System Based on SOC Unit

According the mechanism composition principle based on SOC unit, a systematic theory and method for unified SOC unit modeling of topology, kinematics and dynamics of mechanisms is established (refer to Chap. 7 for detail). The main contents include:

(1) A systematic method for mechanism topology design based on SOC unit [4, 28–46, 59–61, 72, 73].

(2) A modular method for kinematic analysis of mechanisms based on SOC unit. Dimension (number of unknowns) of the kinematic analysis equations can be greatly reduced from the topological structure level using this modular method and is exactly equal to the AKC's coupling degree [61–68].

(3) A modular method for dynamic analysis of mechanisms based on SOC unit. Dimension of the dynamic analysis equations can be greatly reduced from the topological structure level using this modular method. Dimension (number of unknowns) of the inverse dynamic analysis equations is exactly equal to the AKC's coupling degree and dimension (number of unknowns) of the forward analysis dynamic equations is equal to sum of the AKC's coupling degree and DOF of the mechanism [61, 64, 69, 70].

1.3 Summary

(1) Fundamental tasks of robot mechanism topology are:

    (a) to provide a systematic theory and method for mechanism topology design (i.e. invention of new mechanisms).

    (b) to set up new mechanism composition principle and, based on this principle, establish a systematic theory and method for unified modeling of mechanism topology, kinematics and dynamics.

(2) General procedure of mechanism topology design shall include:

    (a) Structure synthesis. Many structure types are obtained.

    (b) Performance analysis, classification and optimization of PM structures.

(3) Four categories of PM topology design methods and their characteristics are briefly introduced.

(4) Four types of mechanism composition principles and their corresponding mechanism theory systems are briefly introduced.

(5) Original work of this book includes:

   (a) Three new basic concepts (geometric constraint type of axes, POC set and SOC unit) are proposed.

   (b) Four basic equations for mechanism topology (POC equation for serial mechanisms, POC equation for parallel mechanisms, DOF formula, AKC coupling degree formula) are proposed.

   (c) 12 topological characteristics of PM are derived. These topological characteristics can be used in performance analysis, classification and optimization of PM structure types.

   (d) A systematic method for topology design of parallel mechanisms is established. The design process includes two stages. The first is structure synthesis in which many different structure types can be obtained. The second is performance analysis, classification and optimization of structure types based on 12 topological characteristics.

   (e) This design method is independent of motion position (excluding singular position) and the fixed coordinate system (i.e. it is not necessary to establish the fixed coordinate system). So, the full-cycle DOF is obtained and geometry conditions of mechanism existence is general. Therefore, this design method is a geometrical method, which is totally different from the other three methods.

   (f) The mechanism composition principle based on SOC unit is proposed. This composition principle provides a theoretical basis for unified modeling of mechanism topology, kinematics and dynamics based on SOC unit.

References

1. Ionescu TG (2003) Terminology for mechanisms and machine science. Mech Mach Theory 38
2. Yang T-L (1983) Structural analysis and number synthesis of spatial mechanisms. In: Proceedings of the 6th world congress, on the theory of machines and mechanisms, New Delhi, vol 1, pp 280–283
3. Yang T-L, Jin Q, Liu A-X, Shen H-P, Luo Y-F (2002) Structural synthesis and classification of the 3-dof translation parallel robot Mechanisms based on the unites of single-open chain. Chin J Mech Eng 38(8):31–36. doi:10.3901/JME.2002.08.031
4. Jin Q, Yang T-L (2004) Theory for topology synthesis of parallel manipulators and its application to three-dimension-translation parallel manipulators. ASME J Mech Des 126:625–639
5. Hunt KH (1978) Kinematic geometry of mechanisms. Clarendon Press, Oxford

6. Frisoli A, Checcacci F et al (2000) Synthesis by screw algebra of translating in-parallel actuated mechanisms. Advance in Robot Kinematics, pp 433–440
7. Huang Z, Li QC (2002) General methodology for type synthesis of symmetrical lower-mobility parallel manipulators and several novel manipulators. Int J Robot Res 21 (2):131–145
8. Huang Z, Li QC (2003) Type synthesis of symmetrical lower-mobility parallel mechanisms using the constraint-synthesis method. Int J Robot Res 22:59–79
9. Kong X, Gosselin CM (2004) Type synthesis of 3T1R 4-dof parallel manipulators based on screw theory. IEEE Trans Rob Autom 20(2):181–190
10. Kong X, Gosselin CM (2007) Type synthesis of parallel mechanisms. Springer Tracts in Advanced Robotics, vol 33
11. Huang Z, Li QC, Ding HF (2012) Theory of parallel mechanisms. Springer, Dordrecht
12. Meng X, Gao F, Wu S, Ge J (2014) Type synthesis of parallel robotic mechanisms: framework and brief review. Mech Mach Theory 78:177–186
13. Herve JM (1978) Analyse structurelle des mécanismes par groupe des déplacements. Mech Mach Theory 13:437–450
14. Hervé JM (1995) Design of parallel manipulators via the displacement group. In: Proceedings of the 9th world congress on the theory of machines and mechanisms Milan, Italy, pp 2079–2082
15. Fanghella P, Galletti C (1995) Metric relations and displacement groups in mechanism and robot kinematics. ASME J Mech Des 117:470–478
16. Selig J (1996) Geometrical methods in robotics. Springer, New York
17. Herve JM (1999) The Lie group of rigid body displacements, a fundamental tool for mechanism design. Mech Mach Theory 34:719–730
18. Li QC, Huang Z, Herve JM (2004) Type synthesis of 3R2T 5-DOF parallel mechanisms using the lie group of displacements. IEEE Trans Rob Autom 20:173–180
19. Li QC, Herve JM (2009) Parallel mechanisms with bifurcation of schonflies motion. IEEE Trans Rob 25(1):158–164
20. Li QC, Herve JM (2010) 1T2R parallel mechanisms without parasitic motion. IEEE Trans Rob 26:401–410
21. Meng J, Liu GF, Li ZX (2007) A geometric theory for analysis and synthesis of sub-6 dof Parallel manipulators. IEEE Trans Rob 23:625–649
22. Rico JM, Cervantes JJ, Rocha J et al (2007) Mobility of single loop linkages: A final word? In: Proceedings of the ASME International Design Engineering Technical Conferences & Computers and Information in Engineering Conference, DETC2007–34936
23. Gogu G (2004) Structural synthesis of fully-isotropic translational parallel robots via theory of linear transformations. Eur J Mech A Solids 23:1021–1039
24. Gogu G (2007) Structural synthesis of fully-Isotropic parallel robots with schonflies motions via theory of linear transformations and evolutionary morphology. Eur J Mech A Solids 26:242–269
25. Gogu G (2008) Structural synthesis of parallel robots: part 1: methodology. Springer, Dordrecht
26. Gogu G (2009) Structural synthesis of parallel robots: part 2: translational topologies with two and three degrees of freedom. Springer, Dordrecht
27. Gogu G (2010) Structural synthesis of parallel robots: part 3: topologies with planar motion of the moving platform. Springer, Dordrecht
28. Gogu G (2012) Structural synthesis of parallel robots: part 4: other topologies with two and three degrees of freedom. Springer, Dordrecht
29. Yang T-L, Jin Q et al (2001) A general method for type synthesis of rank-deficient parallel robot mechanisms based on SOC unit. Mach Sci Technol 20(3):321–325
30. Qiong J (2001) Dissertation: Study on overconstrained mechanisms and low-mobility parallel robot mechanisms. Southeast University, China

31. Yang T-L, Jin Q et al (2001) Structural synthesis of 4-dof (3-translational and 1-rotation) parallel robot mechanisms based on the unites of single-opened-chain. In: The 27th design automation conference, Pittsburgh, ASME International DETC2001/DAC-21152
32. Jin Q, Yang T-L (2001) Structure synthesis of a class five-dof parallel robot mechanisms based on single-opened-chain units. In: The 27th design automation conference, Pittsburgh, ASME International DETC2001/DAC-21153
33. Yang T-L (2004) Theory of topological structure for robot mechanisms. Chin Mach Press, Beijing
34. Jin Q, Yang T-L (2002) Structure Synthesis and analysis of parallel manipulators with 2-dimension translation and 1-dimension rotation. In: Proceedings of ASME 28th design automation conference, Montreal, DETC 2002/MECH 34307
35. Jin Q, Yang T-L (2002) Synthesis and analysis of a group of 3 dof (1T-2R) decoupled parallel manipulator In: Proceedings of ASME 28th design automation conference, Montreal, DETC 2002/MECH 34240
36. Jin Q, Yang T-L (2004) Synthesis and analysis of a group of 3-degree-of-freedom partially decoupled parallel manipulators. ASME J Mech Des 126:301–306
37. Shen H-P, Yang T-L et al (2005) Synthesis and analysis of kinematic structures of 6-dof parallel robotic mechanisms. Mech Mach Theory 40:1164–1180
38. Shen H-P, Yang T-L et al (2005) Structure and displacement analysis of a novel three-translation parallel mechanism. Mech Mach Theory 40:1181–1194
39. Yang T-L, Liu A-X, Luo Y-F et al (2009) Position and orientation characteristic equation for topological design of robot mechanisms. ASME J Mech Des 131:021001-1–021001-17
40. Yang T-L, Liu A-X, Shen H-P et al (2013) On the correctness and strictness of the POC equation for topological structure design of robot mechanisms. ASME J Mech Robot 5 (2):021009-1–021009-18
41. Yang T-L, Sun D-J (2006) General formula of degree of freedom for parallel mechanisms and its application. In: Proceedings of ASME 2006 mechanisms conference, DETC2006-99129
42. Yang T-L, Sun D-J (2008) Rank and mobility of single loop kinematic chains. In: Proceedings of the ASME 32-th mechanisms and robots conference, DETC2008-49076
43. Yang T-L, Sun D-J (2012) A general DOF formula for parallel mechanisms and multi-loop spatial mechanisms. ASME J Mech Robot 4(1):011001-1–011001-17
44. Yang T-L, Liu A-X, Shen H-P et al (2015) Composition principle based on SOC unit and coupling degree of BKC for general spatial mechanisms. In: The 14th IFToMM world congress, Taipei, Taiwan, 25–30 Oct 2015. doi:10.6567/IFToMM.14TH.WC.OS13.135
45. Yang T-L, Shen H-P, Liu A-X, Dai JS (2015) Review of the formulas for degrees of freedom in the past ten years. J Mech Eng 41(9):28–32
46. Yang T-L, Liu A-X, Shen H-P et al (2012) Theory and application of robot mechanism topology. Chin Sci Press, Beijing
47. Yang T-L, Liu A-X, Luo Y-F, Shen H-P et al (2010) Basic principles, main characteristics and development tendency of methods for robot mechanism structure synthesis. J Mech Eng 46(9):1–11
48. Assur LV (1913) Investigation of plane hinged mechanisms with lower pairs from the point of view of their structure and classification (in Russian): Part I, II. Bull Petrograd Polytech Inst 20:329–386. Bull Petrograd Polytech Inst 21(1914):187–283 (vols. 21–23, 1914–1916)
49. Dobrovolskii VV (1939) Main principles of rational classification. AS USSR
50. Mruthyunjaya TS (2003) Kinematic structure of mechanisms revisited. Mech Mach Theor 38:279–320
51. Tuttle ER et al (1989) Enumeration of basic kinematic chains using the theory of finite groups. Trans ASME J Mech Trans Auto Design 111(4):498–503
52. Ceresole E, Fanghella P, Galletti C (1996) Assur's groups, AKCs, basic trusses, SOCs, etc.: modular kinematics of planar linkages. In: Proceeding of 1996 ASME design engineering technical conference 96-DETC/MECH-1027

53. Reuleaux F (1876) Theoretische kinematic Fridrich vieweg, Braunschweig, Germany, 1875 (English translation by A.B.W. Kennedy, The kinematics of machinery, 1876, Reprinted Dover, 1963)
54. Franke R (1951) Von Aufbau der getribe, vol. 1. Beuthvertrieb, Berlin. (1943) vol. 2. VDI, Dusseldorf
55. Hain K (1967) Applied kinematics, 2nd edn. McGraw-Hill, New York
56. Wittenburg J (1977) Dynamics of systems of rigid bodies. Teubner, Stuttgart
57. Paul B (1979) Kinematics and dynamics of planar machinery. Prentice-Hall Inc., New Jersey
58. Sohn WJ, Freudenstein F (1989) An application of dual graphs to the automatic generation of the kinematic structure of mechanisms. ASME J Mech Trans Auto Des 111(4):494–497
59. Tsai L-W (1999) Robot analysis: the mechanics of serial and parallel manipulators. Wiley, New York
60. Yang T-L, Yao F-H (1988) The topological characteristics and automatic generation of structural analysis and synthesis of plane mechanisms, part 1-theory, part 2-application. In: Proceedings of ASME Mechanisms Conference on, Orlando, vol. 1, pp 179–190
61. Yang T-L, Yao F-H (1992) The topological characteristics and automatic generation of structural analysis and synthesis of spatial mechanisms, part 1-topological characteristics of mechanical network; part 2-automatic generation of structure types of kinematic chains. In: Proceedings of ASME Mechanisms Conference, Phoenix, DE-47, pp 179–190
62. Yang T-L (1996) Basic theory of mechanical system: structure, kinematic and dynamic. China Mach Press, Beijing
63. Shen H-P, Yang T-L (1994) A new method and automatic generation for kinematic analysis of complex planar linkages based on the ordered SOC. Proc ASME Mech Conf 70:493–500
64. Kong X, Yang T-L (1995) A zero simple open chain approach to the displacement analysis of multi-loop general spatial linkages. In: Proceedings of the 9th world congress on the theory of machines and mechanisms, Milano, vol 2, 777–781
65. Yang T-L, Yao F-H, Zhang M (1998) A comparative study on some modular approaches for analysis and synthesis of planar linkage. In: Proceedings of ASME mechanisms conference Atlanta, DETC98/MECH-5920
66. Shen H-P, Ting K-L, Yang T-L (2000) Configuration analysis of complex multiloop linkages and manipulators. Mech Mach Theory 35(3):353–362
67. Shi Z-X, Luo Y-F, Yang T-L (2006) Modular method for kinematic analysis of parallel manipulators based on ordered SOCs. In: Proceedings of the ASME 31-th mechanisms and robots conference, DETC2006-99089
68. Shi Z-X, Lao Y-F, Hang L-B, Yang T-L (2007) A simple method for inverse kinematic analysis of the general 6R serial robot. ASME. J Mech Des 129(8):793–798
69. Nicolas R, Federico T (2012) On closed-form solutions to the position analysis of baranov trusses. Mech Mach Theory 50:179–196
70. Zhang J-Q, Yang T-L (1994). A new method and automatic generation for dynamic analysis of complex planar mechanisms based on the SOC. In: Proceedings of 1994 ASME mechanisms conference, vol 71, pp 215–220
71. Yang T-L, Li H-L, Luo Y-F (1991) On the structure of dynamic equation of any mechanical system. Chin J Mech Eng 27(4):1–15
72. Shen C-W, Hang L-B, Yang T-L (2017) Position and orientation characteristics of robot mechanisms based on geometric algebra. Mech Mach Theory 108:231–243
73. Yang T-L, Liu A-X, Shen H-P, Hang L-B (2017) Topological structural synthesis of 3T-1R parallel mechanisms based on POC equations. Chin J Mech Eng 53(21):54–64. doi:10.3901/JME.2017.21.054

Chapter 2
Topological Structure of Mechanisms and Its Symbolic Representation

2.1 Introduction

In order to reveal the mapping relations among topological structure, POC and DOF of mechanisms (i.e. to establish the basic equations for mechanism topology), topological structure of mechanisms and its symbolic representation are discussed.

2.1.1 Traditional Description of Mechanism Topological Structure

In the IFToMM's terminology, the mechanism topological structure is defined as [1]: "**STRUCTURE** (OF A MECHANISM): Number and kinds of elements in a mechanism (members and joints) and the sequence of their contact".

According to this definition, topological structure of a mechanism include: (1) Number and type of pairs (such as P, R, H, C, S pairs, etc.). (2) Number and type of links (such as binary link, ternary link, etc.). (3) Connection relations among these kinematic pairs and links.

The traditional definition involves two key elements of topological structure: type of kinematic pairs and connection relations among links. But these two key elements are not enough to exactly describe topological structure of a mechanism. For the two mechanisms in Fig. 2.4, they have the same description if only these two key elements are used. But actually they have different topological structures.

T.-L. Yang et al., *Topology Design of Robot Mechanisms*,
https://doi.org/10.1007/978-981-10-5532-4_2

2.1.2 Current Description of Mechanism Topological Structure

In this chapter, the following three key elements are used to describe the topological structure of a mechanism:

(1) The first traditional key element: type of kinematic pair.

(2) The second traditional key element: connection relations among links (or topological structure units).

(3) The new key element: type of geometric constraints to pair axes imposed by links (hereafter called as geometric constraint type of axes), refer to Sect. 2.2.3 for detail.

The topological structure of any spatial mechanism can be exactly described using these three key elements of topological structure.

2.1.3 Topology Structure Design of Mechanisms

Topological structure design is to determine the three key elements of a mechanism's topological structure and to draw the schematic diagram of the mechanism.

Topological structure design is hereafter called as topology design. The systematic theory and method for topology design of mechanisms is called as mechanism topology.

2.2 Three Key Elements of Mechanism Topological Structure

2.2.1 Type of Kinematic Pairs

The common used types of kinematic pair include single DOF kinematic pair (such as R, R and H pair) and multi-DOF kinematic pair (such as S and C pair), as shown in Table 2.1.

Table 2.1 Common types of kinematic pairs

P pair	R pair	H pair	C pair	S pair
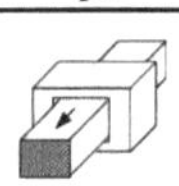	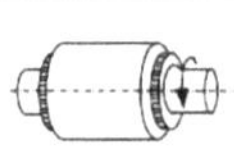	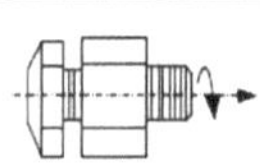	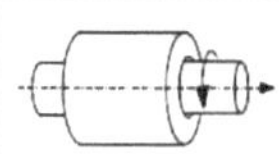	

A kinematic pair brings a certain type of geometric surface constraint to the connected two links. And thus determine the motion characteristics of this kinematic pair. For example, a P pair provides a prismatic surface constraint to the connected two links and thus determines that there is only one translation between these two links. An R pair brings in a cylindrical surface and end surface constraints to the connected two links and thus determines that there is only one rotation between these two links.

2.2.2 SOC Unit and Its Connection Relations

A mechanism can be regarded as being formed by connecting several topological structure units. According to different requirements, the topological structure unit may be of different types, such as link-pair unit, loop unit, Single-Open-Chain (SOC) unit, etc.

The connection relation among topological structure units can be described using different mathematical tools. For example, the connection relation among link-pair unit can be described using adjacency matrix [2–5] and the connection relation among loop units can be described using loop matrix [2, 6, 7].

A single open chain (SOC) refers to a simple open chain by links connected in series by kinematic pairs. Figure 2.1 shows an SOC which is expressed as SOC$\{-R_1 - R_2 - P_3 - \cdots - H_{(j-1)} - R_j-\}$ [8–16]. A sub-SOC refers to a part of an SOC. For example, sub-$SOC\{-R_1 - R_2 - P_3-\}$ and sub-$SOC\{-H_{(j-1)} - R_j-\}$ are two sub-SOCs of the SOC shown in Fig. 2.1.

There are only two types of connection relations between SOCs, serial or parallel. Based on the SOC unit and the two types of connection relations, any serial mechanisms or multi-loop spatial mechanisms can be obtained [10–19]. So, the SOC unit and the connection relations can be used for generating any mechanisms, for establishing the three basic topological structure equations (Eq. (4.4) in Chap. 4, Eq. (5.3) in Chap. 5, and Eq. (6.8) in Chap. 6) and building up a systematic theory and method for unified modelling of mechanism topology, kinematics and dynamics based on the SOC unit (refer to Chap. 7 for detail).

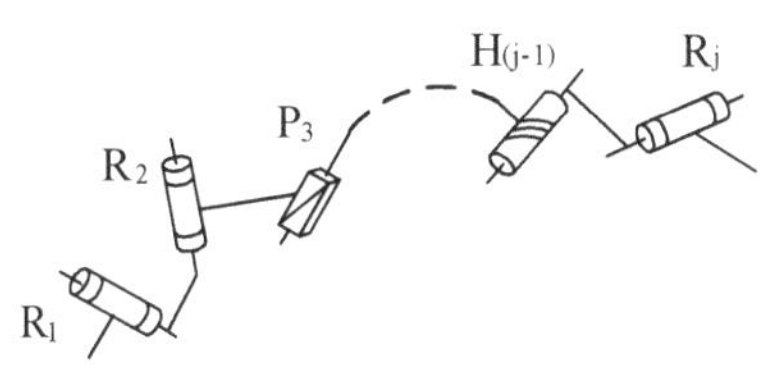

Fig. 2.1 Single open chain (SOC)

2.2.3 Geometric Constraint Type of Axes

Geometric constraint type of axes refers to the type of geometric constraint of a link to the axes of the pairs on this link. In commonly used mechanisms, there are only six types of geometric constraints of axes [8, 11–19]:

(1) Axes of several adjacent R (or H) pairs are parallel, as shown in Fig. 2.2a. It is expressed as $SOC\{-R \parallel R \parallel \cdots \parallel R-\}$. There are many different sub-SOCs with this geometric constraint type of axes, as shown in Table 2.2a.

(2) Axes of two adjacent pairs are coindident, as shown in Fig. 2.2b. It is expressed as $SOC\{-R|H-\}$. There are many different sub-SOCs with this geometric constraint type of axes, as shown in Table 2.2b.

(3) Axes of several adjacent R pairs intersect at one point, as shown in Fig. 2.2c. It is expressed as $SOC\{-\overbrace{RR\cdots R}-\}$. There are many different sub-SOCs with this geometric constraint type of axes, as shown in Table 2.2c.

(4) Axes of two adjacent pairs are perpendicular, as shown in Fig. 2.2d. It is expressed as $SOC\{-R\perp P-\}$. There are many different sub-SOCs with this geometric constraint type of axes, as shown in Table 2.2d.

(5) Axes of several P pairs are parallel to one plane, as shown in Fig. 2.2e. It is expressed as $SOC\{-\diamondsuit(P,P,\cdots,P-)\}$. There are many different sub-SOCs with this geometric constraint type of axes, as shown in Table 2.2e.

(6) Axes of kinematic pairs are allocated arbitrarily in the space, as shown in Fig. 2.2f. It is expressed as $SOC\{-R-R-\}$. There are many different sub-SOCs with this geometric constraint type of axes, as shown in Table 2.2f.

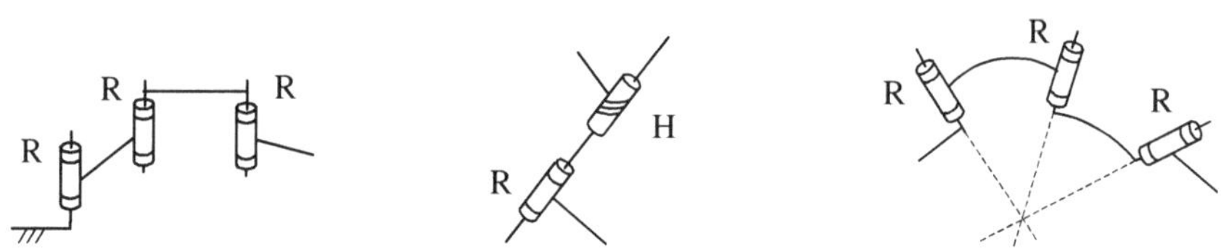

(a) $SOC\{-R \parallel R \parallel \cdots \parallel R-\}$ (b) $SOC\{-R|H-\}$ (c) $SOC\{-\overbrace{RR\cdots R}-\}$

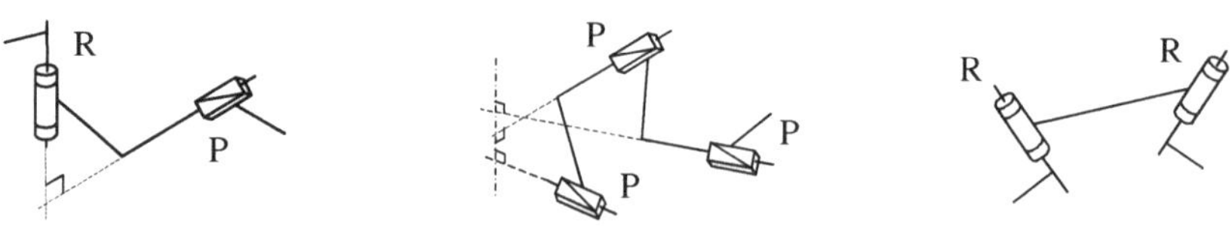

(d) $SOC\{-R\perp P-\}$ (e) $SOC\{-\diamondsuit(P,P,\cdots,P-)\}$ (f) $SOC\{-R-R-\}$

Fig. 2.2 Geometric constraint types of axes [11–15]

Table 2.2 Sub-SOCs of geometric constraint types

Geometric constraint type	Sub-SOCs
(a) Parallel to each other	$R \parallel R$, $R \parallel R \parallel R$, $R \parallel H$, $H \parallel H$, $H \parallel H \parallel H$, etc.
(b) Coincident axes	$R\|R$, $R\|P$, $H\|R$, $H\|H$, $H\|P$, etc.
(c) Intersecting at one point	$\overset{\frown}{RRR}$, $\overset{\frown}{RR}$, $\overset{\frown}{R\perp R}$, etc.
(d) Perpendicular to each other	$R\perp R$, $R\perp P$, $H\perp P$, $P\perp P$, etc.
(e) Parallel to a plane	$\diamondsuit(P, P, P)$, etc.
(*f*) Arbitrary	$R-R$, $R-P$, $R-H$, $H-H$, etc.

Note The special geometric relations among pair axes are first proposed by Dizioglu [20]. We concluded these special geometric relations to six geometric constraint types in 1983 and applied them in structure synthesis of serial mechanisms and overconstrained loops [8, 9]. In 1992, these geometric constraint types were used in structure synthesis of multi-loop spatial mechanisms [10]. In 2001, the geometric constraint type was regarded as one of the three key elements to describe the topological structure of mechanisms [11–15]. Now, the geometric constraint types are widely used in structure analysis and synthesis of robot mechanisms [11–19]

2.3 Symbolic Representation of Mechanism Topological Structure

2.3.1 Symbolic Representation of Multi-DOF Pair

A multi-DOF pair can be regarded as being formed by several single-DOF pairs connected in series. So, a multi-DOF pair can be expressed as an SOC composed of several single-DOF pairs. For example, the S pair can be expressed as $SOC\{-\overset{\frown}{RRR}-\}$, the C pair can be expressed as $SOC\{-R|P-\}$, the E pair can be expressed as $SOC\{-P\perp R\perp P-\}$ and the U pair can be expressed as $SOC\{-\overset{\frown}{R\perp R}-\}$.

2.3.2 Symbolic Representation of Topological Structure

Based on the three key elements of mechanism topological structure, the topological structure of any spatial mechanism can be exactly described using three types of pairs (P, R and H), six geometric constraint types of axes and two connection relations between SOC units (serial or parallel) and can be expressed symbolically.

For example, topological structure of the serial mechanism in Fig. 2.3a can be expressed symbolically as $SOC\{-R_1(\perp P_2) \parallel R_3 - \overset{\frown}{R_4R_5}-\}$. In this symbolic expression, $R_1(\perp P_2) \parallel R_3$ means $R_1 \perp P_2$ and $R_1 \parallel R_3$.

The second example is the parallel mechanism shown in Fig. 2.3b. Topological structure of this parallel mechanism can be expressed symbolically as follows:

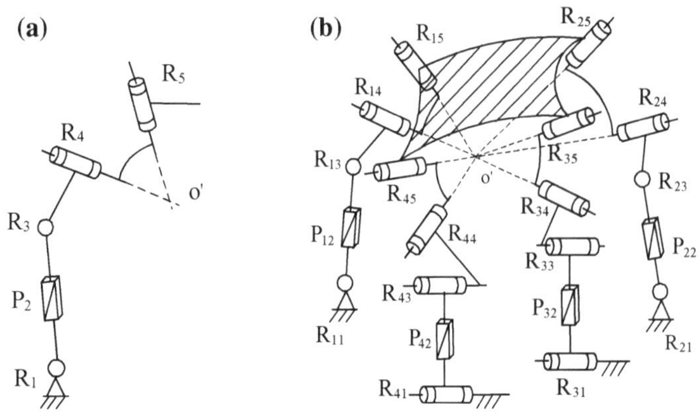

Fig. 2.3 Mechanism topological structures

(a) Topological structure of four branches: $4 - SOC\{-R_{i1}(\perp P_{i2}) \parallel R_{i3} - \widehat{R_{i4}R_{i5}} -\}$(i = 1–4).

(b) Topological structure of the moving platform: R_{i4}, R_{i5} (i = 1–4) intersect at point o′.

(c) Topological structure of the fixed platform: $R_{11} \parallel R_{21}, R_{31} \parallel R_{41}$, R_{11} and R_{31} are skewed arbitrarily.

Another example includes the two single-loop chains in Fig. 2.4. The single-loop chain in Fig. 2.4a can be expressed symbolically as $SLC\{\widehat{-R_1R_2R_3} - \widehat{R_4R_5R_6 -}\}$ and the single-loop chain in Fig. 2.4b can be expressed symbolically as $SLC\{-R_1 \parallel R_2 \parallel R_3 - \widehat{R_4R_5R_6} -\}$. From this example, we find that it is necessary to use geometric constraint type of axes to describe mechanism topological structure. Otherwise, it will be very hard for us to show the different topological structures of these two single-loop chains since they contain the same type and number of pairs.

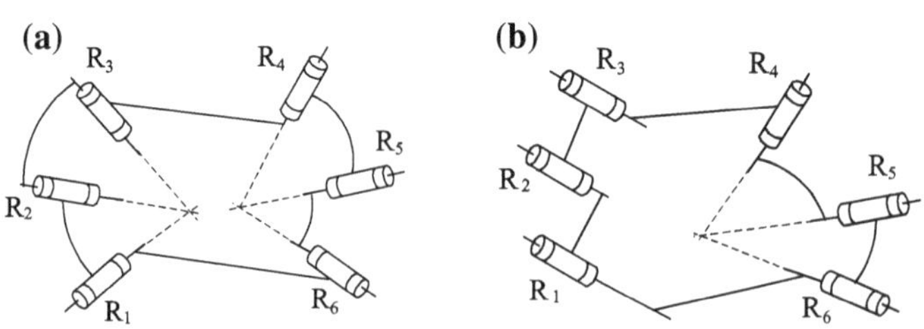

Fig. 2.4 Topological structure of single-loop chains

2.3.3 Composition of an SOC

(1) An SOC can be formed by connecting several pairs in Table 2.1 in series, as shown in Fig. 2.1.

(2) An SOC can be formed by connecting several sub-SOCs in Table 2.2 in series. For example, $SOC\{-R \parallel R \parallel R - \widehat{RR} -\}$ can be regarded as being formed by connecting sub-$SOC\{-R \parallel R \parallel R-\}$ and sub-$SOC\{- \widehat{RR} -\}$(refer to Table 2.2) in series.

(3) An SOC can be formed by connecting several SOCs in series. For example, $SOC\{-R_{11}(\perp P_{12}) \parallel R_{13} - \widehat{R_{14}R_{15}R_{25}R_{24}} -R_{23}(\perp P_{22}) \parallel R_{21}-\}$ corresponding to first loop of the parallel mechanism (Fig. 2.3b) is formed by connecting $SOC\{-R_{i1}(\perp P_{i2}) \parallel R_{i3} - \widehat{R_{i4}R_{i5}} -\}$ (i = 1, 2) corresponding to the first two branches in series.

2.4 Topological Structure Invariance

Generally, topological structure of a mechanism is determined by design, processing and assembling. Topological structure described by the three key elements is invariant during motion process of the mechanism (excluding singular positions). In other words, topological structure of a mechanism is independent of motion position of the mechanism and the fixed coordinate system (i.e. it is not necessary to establish the fixed coordinate system).

For example, during motion process of the mechanism in Fig. 2.5a, the singular position refers to such a position that axes of non-adjacent kinematic pairs (R_1, R_2, R_4 and R_5) intersect temporarily at one point $o_1(o_2)$, as shown in Fig. 2.5b. In other nonsingular positions, the topological structure is invariant, as shown in Fig. 2.5a and Fig. 2.6.

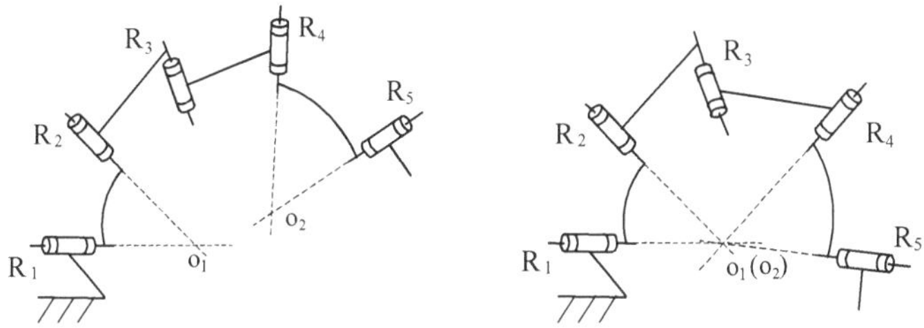

(a) Non-singular position (b) Singular position

Fig. 2.5 Singular position of non-adjacent pair axes

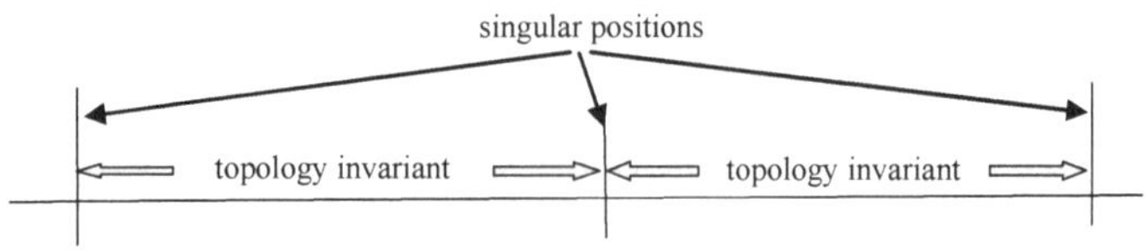

Fig. 2.6 Invariance property of mechanism topological structure

Correspondingly, symbolic representation of the mechanism topological structure based on three key elements is also motion process invariant (excluding singular positions). So this symbolic representation is a kind of geometric representation [11–19].

The topological structure invariance can be used for establishing the three basic topological structure equations (Eq. (4.4) in Chap. 4, Eq. (5.3) in Chap. 5 and Eq. (6.8) in Chap. 6).

2.5 Definition of Mechanism Topological Structure

In order to realize prescribed functions and performance requirements, pairs and links of a mechanism must be connected in a certain way. This connection way is called topological structure of the mechanism. In the topological structure, the size of pairs or links and the space between them are not considered. But the geometric constraint type of axes must be kept unchanged [19].

2.6 Classification of Mechanisms Based on Topological Structure

According to definition of the topological structure in Sect. 2.5, there are the following three types of mechanisms:

(1) Non-overconstrained mechanisms

A non-overconstrained mechanism refers to such a mechanism that number of its independent displacement equations is six and the DOF ≥ 1.

Existence condition of non-overconstrained mechanism is related only to types of pairs and connection relations among links. The non-overconstrained mechanism is also called trivial kinematic chain [21].

For example, the general 7R mechanism and the 6-SPS parallel mechanism are non-overconstrained mechanisms.

(2) General overconstrained mechanism

A general overconstrained mechanism refers to such a mechanism that number of its independent displacement equations is less than six and the DOF ≥ 1.

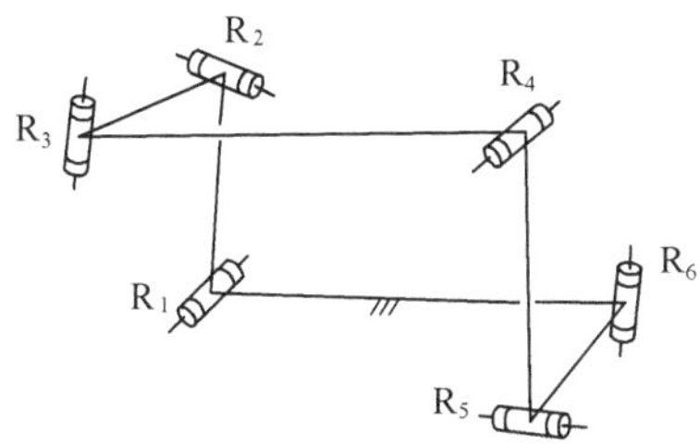

Fig. 2.7 Bricard mechanism

Existence condition of general overconstrained mechanism is related to types of pairs, connection relations among links and the geometric constraint types of axes. The general overconstrained mechanism is also called exceptional kinematic chain [21].

For example, the mechanisms shown in Figs. 2.3b and 2.4 are general overconstrained mechanisms.

(3) Special overconstrained mechanism

A special overconstrained mechanism refers to such a mechanism that number of its independent displacement equations is less than six and the DOF = 1.

Existence condition of special overconstrained mechanism is related to types of pairs, connection relations among links and a certain functional relationship with link parameters (link length and skew angle comply with a certain functional relationship). The special overconstrained mechanism is also called paradoxical kinematic chain [21].

For example, Bricard mechanism is a six-bar special overconstrained mechanism [22, 23], as shown in Fig. 2.7. Its existence conditions include: $d_i = 0$ (i = 1–6), $\alpha_i = \pi/2$ (i = 1–6) and $a_1^2 + a_3^2 + a_5^2 = a_2^2 + a_4^2 + a_6^2$.

The systematic theory and method for topology design covers only non-overconstrained mechanisms (the trivial kinematic chains) and general overconstrained mechanisms (the exceptional kinematic chains). Researches on special overconstrained mechanisms (the paradoxical kinematic chains) may be found in other references [22–24].

2.7 Summary

(1) Mechanism topological structure contains three key elements: types of pairs (Table 2.1), connection relations among topological structure units (such as link-pair unit, loop unit and SOC unit) and geometric constraint types of axes (Fig. 2.2). The topological structure of any spatial mechanism can be exactly described using these three elements.

(2) Any serial mechanisms or multi-loop spatial mechanisms can be generated by connecting SOC units parallels or in series.

(3) Mechanism topological structure has invariance property and is independent of motion position of the mechanism (excluding singular positions) and the fixed coordinate system (i.e. it is not necessary to establish the fixed coordinate system) (Fig. 2.6).

(4) Symbolic representation of mechanism topological structure is a geometric expression and is independent of motion position of the mechanism and the fixed coordinate system. So, it provides a theoretical basis for operations of mechanism topology (refer to Chaps. 3–5) for detail.

(5) According to definition of mechanism topological structure in Sect. 2.5, there are three types of mechanisms: non-overconstrained mechanisms (trivial kinematic chains), general overconstrained mechanisms (exceptional kinematic chains) and special overconstrained mechanisms (paradoxical kinematic chains).

References

1. Ionescu TG (2003) Terminology for mechanisms and machine science. Mech Mach Theor 38
2. Mruthyunjaya TS (2003) Kinematic structure of mechanisms revisited. Mech Mach Theor 38:279–320
3. Franke R (1951) Von Aufbau der Getribe, vol 1, Beuthvertrieb, Berlin, 1943, vol 2, VDI, Dusseldorf
4. Hain K (1967) Applied Kinematics, 2nd edn. McGraw-Hill, New York
5. Wittenburg J (1977) Dynamics of systems of rigid bodies. Teubner, Stuttgart
6. Paul B (1979) Kinematics and dynamics of planar machinery. Prentice-Hall Inc, New Jersey
7. Sohn WJ, Freudenstein F (1989) An application of dual graphs to the automatic generation of the kinematic structure of mechanisms. ASME J Mech Trans Autom Des 111(4):494–497
8. Yang T-L (1983) Structural analysis and number synthesis of spatial mechanisms. In: Proceedings of the 6th world congress on the theory of machines and mechanisms, New Delhi, vol 1, pp 280–283
9. Yang T-L (1986) Kinematic structural analysis and synthesis of over-constrained spatial single loop chains. In: Proceedings of the 19-th Biennial Mechanisms Conference, Columbus, ASME paper 86-DET-189
10. Yang T-L, Yao F-H (1992) The topological characteristics and automatic generation of structural analysis and synthesis of spatial mechanisms. In: Proceedings of ASME Mechanisms Conference, Phoenix, DE-47, pp 179–90
11. Yang T-L, Jin Q et al (2001) A general method for type synthesis of rank-deficient parallel robot mechanisms based on SOC unit. Mach Sci Technol 20(3):321–325
12. Yang T-L, Jin Q, Liu A-X, Shen H-P, Luo Y-F (2002) Structural synthesis and classification of the 3-DOF translation parallel robot mechanisms based on the unites of single-open chain. Chin J Mech Eng 38(8):31–36. doi:10.3901/JME.2002.08.031
13. Jin Q, Yang T-L (2004) Theory for topology synthesis of parallel manipulators and its application to three-dimension-translation parallel manipulators. ASME J Mech Des 126:625–639
14. Yang T-L (2004) Theory of topological structure for robot mechanisms. China Machine Press, Beijing

15. Yang T-L, Liu A-X, Luo Y-F, Shen H-P et al (2009) Position and orientation characteristic equation for topological design of robot mechanisms. ASME J Mech Des 131: 021001-1–021001-17
16. Yang T-L, Sun D-J (2012) A general DOF formula for parallel mechanisms and multi-loop spatial mechanisms. ASME J Mech Robot 4(1):011001-1–011001-17
17. Yang T-L, Liu A-X, Luo Y-F, Shen H-P et al (2013) On the correctness and strictness of the position and orientation characteristic equation for topological design of robot mechanisms. ASME J Mech Robot 5:021009-1–021009-18
18. Yang T-L, Liu A-X, Shen H-P, et al (2015) Composition principle based on SOC unit and coupling degree of BKC for general spatial mechanisms. In: The 14th IFToMM world congress, Taipei, Taiwan, 25–30 Oct 2015. doi: 10.6567/IFToMM.14TH.WC.OS13.135
19. Yang T-L, Liu A-X, Shen H-P et al (2012) Theory and application of robot mechanism topology. China Science press, Beijing
20. Dizioglu B (1978) Theory and practice of spatial mechanisms with special positions of the axes. Mech Mach Theor 13(2):139–153
21. Herve JM (1978) Analyse Structurelle des Meanismes Dar Groupe des delacements. Mech Mach Theor 4:437–450
22. Baker JE (1979) The Bennett, Goldberg and Myard linkages—in perspective. Mech Mach Theor 14:239–253
23. Baker JE (1980) An analysis of the bricard linkages. Mech Mach Theor 15:267–286
24. Jin Q, Yang T-L (2002) Over-constraint analysis on spatial 6-link loops. Mech Mach Theor 37(3):267–278

Chapter 3
Position and Orientation Characteristics Set

3.1 Introduction

In order to reveal the mapping relations among mechanism topological structure, POC and DOF (i.e. to establish the basic equations for mechanism topology), relative motion characteristics between any two links of a mechanism and corresponding mathematical representation are discussed.

Supposing coordinate system on link i is o′-x′y′z′ (o′ is the origin or base point) and coordinate system on link j is o-xyz, the relative motion characteristics of link i relatively to link j can be described with three rotation elements and three translation elements. Up to now, there are four types of methods for description of relative motion characteristics between links of a mechanism.

(1) Method based on screw theory

With this method, six components of the screw's Plucker coordinate are used to describe relative motion characteristics between two links and used in structure synthesis of parallel mechanisms [1–4]. It is dependent on motion position of the mechanism and the fixed coordinate system.

(2) Method based on displacement subgroup/submanifold

With this method, 12 kinds of displacement subgroups are used for describing relative motion characteristics of a mechanism fulfilling the algebra structures of Lie group, which is independent of motion position of the mechanism [4–6]. But, for mechanisms not fulfilling the algebra structures of Lie group, the displacement submanifold based method could be used for their structure analysis and synthesis, which depends on motion position of the mechanism [4, 7].

T.-L. Yang et al., *Topology Design of Robot Mechanisms*,
https://doi.org/10.1007/978-981-10-5532-4_3

(3) Method based on theory of linear transformations.

With this method, velocity space is used to describe relative motion characteristics between two links and used in structure synthesis of parallel mechanisms [4, 8]. It is dependent on motion position of the mechanism and the fixed coordinate system.

(4) Method based on POC set

Six elements of POC set are used to describe relative motion characteristics between two links. Three elements are used to describe the position and orientation characteristics of relative rotation. The other three elements are used to describe the position and orientation characteristics of relative translation. It is independent of motion position (excluding singular position) of the mechanism and the fixed coordinate system (i.e. it is not necessary to establish the fixed coordinate system). The POC equations (Eqs. 4.4 and 5.3) are established based on this method [4, 9–17].

The POC set is the core concept of this chapter. The basic idea of POC set is as follows: firstly, derive VC set based on the unit vector set of link velocity of a mechanism. Then derive POC set based on the VC set invariance. Both VC set and POC set are independent of motion position of the mechanism (excluding singular positions) and the fixed coordinate system.

3.2 POC Sets of Kinematic Pairs

3.2.1 POC Set of a P Pair

(1) Velocity analysis of a P pair

As shown in Fig. 3.1, velocity of the moving link is

$$\begin{cases} \omega_P \equiv 0 \\ v_{o'} = |v_{o'}| e_P \end{cases}$$

where, ω_P—angular velocity, $v_{o'}$—translational velocity of the base point o′ on the moving link (origin of the moving coordinate system), e_P—unit vector in direction of the translational velocity, as shown in Fig. 3.1a.

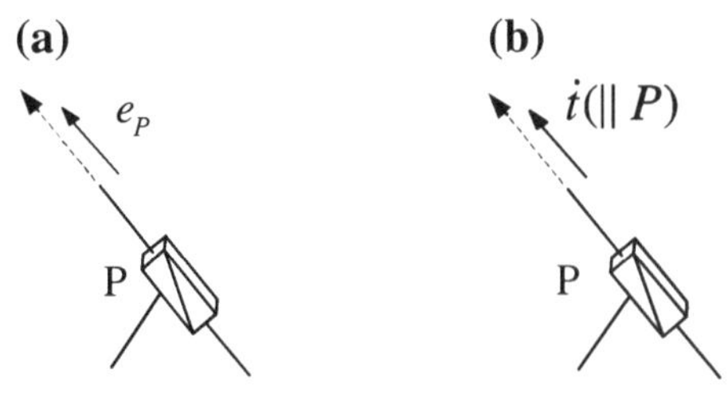

Fig. 3.1 Velocity analysis of a P pair

(2) VC set of a P pair

According to above velocity analysis, velocity of the moving link is related to topological structure, dimensional parameters and motion inputs. But unit vector of the velocity component is independent of dimensional parameters or motion inputs. Therefore, the unit vector can be used to describe velocity characteristics of a P pair as follows:

(a) $\omega_P = 0$ means that there exists no angular velocity. It can be expressed as $\dot{r} \equiv 0$.

(b) Unit vector e_P can be rewritten as $\dot{t}^1(\| P)$. It means that there exists a translational velocity along direction of the P pair, as shown in Fig. 3.1b.

Thus, VC set of a P pair is defined as

$$\dot{M}_P = \begin{bmatrix} \dot{t}^1(\| P) \\ \dot{r} \equiv 0 \end{bmatrix}$$

where, $\dot{M}_P$—VC set of a P pair, $\dot{r} \equiv 0$—there exists no angular velocity, $\dot{t}^1(\| P)$—there exists a translational velocity for the base point o′ along direction of the P pair.

Velocity characteristics of a P pair depend only on structure of the pair, so the VC set is invariant during the motion process.

(3) POC set of a P pair

According to invariance property of the VC set of a P pair, component of the VC set also features invariance. The element $\dot{r} \equiv 0$ means that there exists no finite rotation and it is written as r^0. Similarly, The element $\dot{t}^1(\| P)$ means that there exists a finite translation along direction of the P pair and it is written as $t^1(\| P)$.

Therefore, POC set of a P pair is defined as [9–13]

$$M_P = \begin{bmatrix} t^1(\| P) \\ r^0 \end{bmatrix} \tag{3.1}$$

where, M_P—POC set of a P pair, r^0—there exists no finite rotation, $t^1(\| P)$—there exists a finite translation along direction of the P pair.

According to invariance property of the VC set, POC set of a P pair also features invariance property.

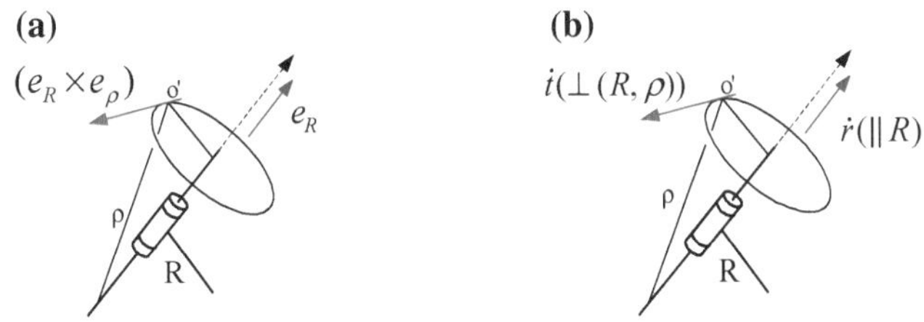

Fig. 3.2 Velocity analysis of an R pair

3.2.2 POC Set of an R Pair

(1) Velocity analysis of an R pair

As shown in Fig. 3.2, velocity of the moving link is

$$\begin{cases} \omega_R = |\omega_R| e_R \\ v_{o'} = |\omega_R| \cdot |\rho| (e_R \times e_\rho) \end{cases}$$

where, ω_R—angular velocity, e_R—unit vector in direction of R pair axis, as shown in Fig. 3.2a, $v_{o'}$—translational velocity of the base point o′ on the moving link, ρ—radius vector from a point on the axis to based point o′, e_ρ—unit vector of the radius vector, as shown in Fig. 3.2a.

(2) VC set of an R pair

According to above velocity analysis, unit vector of velocity can be used to describe velocity characteristics of an R pair.

(a) Unit vector of angular velocity e_R can be rewritten as $\dot{r}^1(\| R)$. It means that there exists an angular velocity parallel to axis of the R pair, as shown in Fig. 3.2b.

(b) Unit vector of translational velocity $(e_R \times e_\rho)$ can be rewritten as $\dot{t}^1(\perp(R, \rho))$. It means that there exists a translational velocity along the direction perpendicular to the plane formed by axis of the R pair and the radius vector ρ, as shown in Fig. 3.2 b. But this translational velocity characteristic and the angular velocity characteristic are dependent.

Thus, VC set of an R pair is defined as

$$\dot{M}_R = \begin{bmatrix} \dot{t}^1(\perp(R, \rho)) \\ \dot{r}^1(\| R) \end{bmatrix}$$

where, $\dot{M}_R$—VC set of an R pair, $\dot{r}^1(\| R)$—there exists an angular velocity parallel to axis of the R pair, $\dot{t}^1(\perp(R, \rho))$—there exists a translational velocity for the base point o′ along the direction perpendicular to the plane formed by axis of the R pair and the radius vector ρ.

Velocity characteristics of an R pair depend only on structure of the pair, so the VC set is invariant during the motion process.

(3) POC set of an R pair

According to invariance property of the VC set of an R pair, component of the VC set also features invariance property. The element $\dot{r}^1(\| R)$ means that there exists a finite rotation parallel to axis of the R pair and it is written as $r^1(\| R)$. Similarly, The element $\dot{t}^1(\perp(R, \rho))$ means that there exists a finite translation for the base point o′ along the direction perpendicular to the plane formed by axis of the R pair and the radius vector ρ and it is written as $t^1(\perp(R,\rho))$.

Therefore, POC set of an R pair is defined as [9–13]

$$M_R = \begin{bmatrix} t^1(\perp(R,\rho) \\ r^1(\| R) \end{bmatrix} \tag{3.2}$$

where, M_R—POC set of an R pair, $r^1(\| R)$—there exists a finite rotation parallel to axis of the R pair, $t^1(\perp(R,\rho))$—there exists a finite translation for the base point o′ along the direction perpendicular to the plane formed by axis of the R pair and the radius vector ρ.

According to invariance property of the VC set, POC set of an R pair also features invariance property.

POC set of an R pair (Eq. 3.2) has two elements, but DOF of an R pair is DOF = 1. So only one element is independent and the other element is a non-independent derivative element. Therefore, POC set of an R pair has two different forms:

(a) $r^1(\| R)$ is selected to be the independent element. $\{t^1(\perp(R,\rho))\}$ is the non-independent derivative element and is put in a brace (R-No. 1 in Table 3.1).

(b) $t^1(\perp(R,\rho))$ is selected to be the independent element. $\{r^1(\| R)\}$ is the non-independent derivative element and is put in a brace (R-No. 2 in Table 3.1).

Hence, POC set of an R pair features duplicity.

Table 3.1 POC sets of kinematic pairs [9–13]

P pair	R pair	H pair
No. 1	R-No. 1	H-No. 1
$\begin{bmatrix} t^1(\| P) \\ r^0 \end{bmatrix}$	$\begin{bmatrix} \{t^1(\perp(R,\rho)\} \\ r^1(\| R) \end{bmatrix}$	$\begin{bmatrix} \{t^1(\| H)\}\ \{t^1(\perp(H,\rho))\} \\ r^1(\| H) \end{bmatrix}$
	R-No. 2	H-No. 2
	$\begin{bmatrix} t^1(\perp(R,\rho) \\ \{r^1(\| R)\} \end{bmatrix}$	$\begin{bmatrix} t^1(\| H)\ \{t^1(\perp(H,\rho))\} \\ \{r^1(\| H)\} \end{bmatrix}$
		H-No. 3
		$\begin{bmatrix} \{t^1(\| H)\}\ t^1(\perp(H,\rho)) \\ \{r^1(\| H)\} \end{bmatrix}$

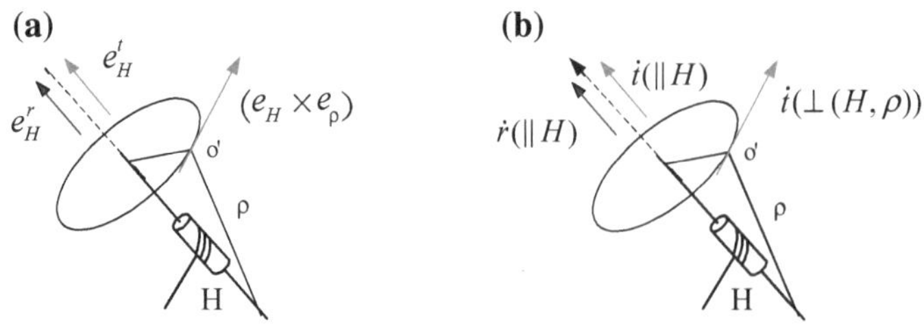

Fig. 3.3 Velocity analysis of an H pair

3.2.3 *POC Set of an H Pair*

(1) Velocity analysis of an H pair

As shown in Fig. 3.3, velocity of the moving link is

$$\begin{cases} \omega_H = |\omega_H| e_H \\ v_{0'} = |\omega_H| \cdot |\rho| (e_H \times e_\rho) + |\omega_H| h e_H \end{cases}$$

where, ω_H—angular velocity, e_H—unit vector in direction of H pair axis, as shown in Fig. 3.3a, $v_{o'}$—translational velocity of the base point o′ on the moving link, ρ—radius vector from a point on the axis to based point o′, e_ρ—unit vector of the radius vector, as shown in Fig. 3.3a, h—thread pitch.

(2) VC set of an H pair

According to above velocity analysis, unit vector of velocity can be used to describe velocity characteristics of an H pair.

(a) Unit vector of angular velocity e_H^r can be rewritten as $\dot{r}^1(\| H)$. It means that there exists an angular velocity parallel to axis of the H pair, as shown in Fig. 3.3b.

(b) Unit vector of translational velocity $(e_H \times e_\rho)$ can be rewritten as $\dot{t}^1(\perp(H,\rho))$. It means that there exists a translational velocity along the direction perpendicular to the plane formed by axis of the H pair and the radius vector ρ, as shown in Fig. 3.3b.

(c) Unit vector of translational velocity e_H^t can be rewritten as $\dot{t}^1(\| H)$. It means that there exists a translational velocity parallel to axis of the H pair, as shown in Fig. 3.3b.

Thus, VC set of an H pair is defined as

$$\dot{M}_H = \begin{bmatrix} \dot{t}^1(\| H) \quad \dot{t}^1(\perp(H,\rho)) \\ \dot{r}^1(\| H) \end{bmatrix}$$

where, $\dot{M}_H$—VC set of an H pair, $\dot{r}^1(\| H)$—there exists an angular velocity parallel to axis of the H pair, $\dot{t}^1(\| H)$—there exists a translational velocity for the base point o′ along axis of the H pair, $\dot{t}^1(\perp(H,\rho))$—there exists a translational velocity for the base point o′ along the direction perpendicular to the plane formed by axis of the R pair and the radius vector ρ.

Velocity characteristics of an H pair depend only on structure of the pair, so the VC set is invariant during the motion process.

(3) POC set of an H pair

According to invariance property of the VC set of an R pair, component of the VC set also features invariance property. The element $\dot{r}^1(\| H)$ means that there exists a finite rotation parallel to axis of the H pair and it is written as $r^1(\| H)$. Similarly, The element $\dot{t}^1(\perp(H,\rho))$ means that there exists a finite translation for the base point o′ along the direction perpendicular to the plane formed by axis of the H pair and the radius vector ρ and it is written as $t^1(\perp(H,\rho))$. The element $\dot{t}^1(\| H)$ means that there exists a finite translation parallel to axis of the H pair and it is written as $t^1(\| H)$.

Therefore, POC set of an H pair is defined as [9–13]

$$M_H = \begin{bmatrix} t^1(\| H) & t^1(\perp(H,\rho)) \\ & r^1(\| H) \end{bmatrix} \tag{3.3}$$

where, M_H—POC set of an H pair, $r^1(\| H)$—there exists a finite rotation parallel to axis of the H pair, $t^1(\| H)$—there exists a finite translation for the base point o′ parallel to axis of the H pair, $t^1(\perp(H,\rho))$—there exists a finite translation for the base point o′ along the direction perpendicular to the plane formed by axis of the R pair and the radius vector ρ.

According to invariance property of the VC set, POC set of an H pair also features invariance property.

POC set of an H pair (Eq. 3.3) has three elements, but DOF of an H pair is DOF = 1. So only one element is independent and the other two elements are non-independent derivative elements. Therefore, POC set of an H pair has three different forms:

(a) $r^1(\| H)$ is selected to be the independent element. The other two elements are non-independent derivative elements and are put in braces (H-No. 1 in Table 3.1).

(b) $t^1(\| H)$ is selected to be the independent element. The other two elements are non-independent derivative elements and are put in braces (H-No. 2 in Table 3.1).

(c) $t^1(\perp(H,\rho))$ is selected to be the independent element. The other two elements are non-independent derivative elements and are put in braces (H-No. 3 in Table 3.1).

Hence, POC set of an H pair features triplicate.

3.3 POC Set of a Mechanism

3.3.1 An Introductory Example

(1) End link velocity analysis of the planar serial mechanism in Fig. 3.4

For the serial mechanism ($SOC\{-R_1 \parallel R_2-\}$) shown in Fig. 3.4, select point o′ on link 2 as the base point (i.e. origin of the moving coordinate system established on link 2). According to principle of motion synthesis, velocity of the end link relative to the frame link is

$$\begin{cases} \omega = |\omega_{1,0}|e_{R_1} + |\omega_{2,1}|e_{R_2} = (|\omega_{1,0}| + |\omega_{2,1}|)e_{R_2} \\ v_{o'} = |\omega_{1,0}| \cdot |\rho_1|(e_{R_1} \times e_{\rho_1}) + |\omega_{2,1}| \cdot |\rho_2|(e_{R_2} \times e_{\rho_2}) \end{cases}$$

where, ω—angular velocity of the end link relative to the frame link, $v_{o'}$—translational velocity of base point o′ on end link relative to the frame link, e_{R_1}—unit vector in direction of the R_1 axis, e_{R_2}—unit vector in direction of the R_2 axis, $\omega_{1,0}$—relative angular velocity between the two links connected by R_1, $\omega_{2,1}$—relative angular velocity between the two links connected by R_2, ρ_1—radius vector from R_1 center to base point o′ and its unit vector is e_{ρ_1}, ρ_2—radius vector from R_2 center to base point o′ and its unit vector is e_{ρ_2}, $|\omega_{1,0}| \cdot |\rho_1|(e_{R_1} \times e_{\rho_1})$—derivative translational velocity of base point o′ generated by the relative angular velocity $\omega_{1,0}$ and its unit vector is $(e_{R_1} \times e_{\rho_1})$, $|\omega_{2,1}| \cdot |\rho_2|(e_{R_2} \times e_{\rho_2})$—derivative translational velocity of base point o′ generated by the relative angular velocity $\omega_{2,1}$ and its unit vector is $(e_{R_2} \times e_{\rho_2})$.

Velocity of the end link can be expressed in the velocity component set:

$$V_S = \begin{bmatrix} |\omega_{1,0}| \cdot |\rho_1|(e_{R_1} \times e_{\rho_1}) & |\omega_{2,1}| \cdot |\rho_2|(e_{R_2} \times e_{\rho_2}) & 0 \\ 0 & 0 & (|\omega_{1,0}| + |\omega_{2,1}|)e_{R_2} \end{bmatrix}$$

The upper row in the set represents translational velocity components in different directions. The lower row shows angular velocity components around different axes. This set contains one angular velocity component of the end link and two translational velocity components of the based point o′. These components are linear combinations of velocity components of all kinematic pairs.

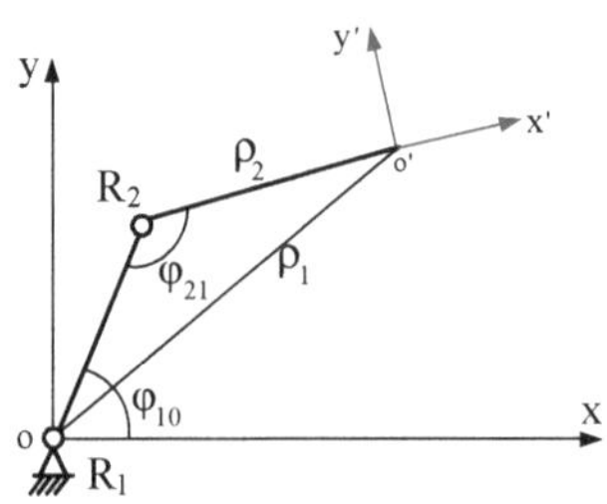

Fig. 3.4 A planar serial mechanism ($SOC\{-R_1 \parallel R_2-\}$)

According to the velocity analysis, velocity of the end link depends on the mechanism topological structure, dimensional parameters, motion position and motion inputs. But unit vector of velocity depends only on the mechanism topological structure and is independent of the dimensional parameters, motion position and motion inputs. Therefore, unit vector set of the velocity can be used to describe the relative velocity characteristics of the end link, i.e.

$$V_S^e = \begin{bmatrix} (e_{R_1} \times e_{\rho_1}) & (e_{R_2} \times e_{\rho_2}) & 0 \\ 0 & 0 & e_{R_2} \end{bmatrix} \tag{3.4}$$

Obviously, unit vector set of the end link velocity is "Union" of unit vector set of each pair velocity. It depends only on topological structure of the mechanism and is independent of motion position (excluding singular position) of the mechanism and the fixed coordinate system (i.e. it is not necessary to establish the fixed coordinate system). Since topological structure of a mechanism has the invariance (refer to the Chap. 2), the unit vector set also features invariance.

(2) VC set of the end link

The unit vector in Eq. (3.4) can be expressed in the form of velocity characteristics as follows:

(a) Unit vector e_{R_2} can be rewritten as $\dot{r}^1(\| R_2)$. It means that the end link has an angular velocity parallel to axis of the R_2 pair.

(b) Unit vector of translational velocity $(e_{R_1} \times e_{\rho_1})$ can be rewritten as $\dot{t}^1(\perp(R_1, \rho_1))$. It means that the base point o' on the end link has a translational velocity along the direction perpendicular to the plane formed by axis of the R_1 and the radius vector ρ_1.

(c) Unit vector of translational velocity $(e_{R_2} \times e_{\rho_2})$ can be rewritten as $\dot{t}^1(\perp(R_2, \rho_2))$. It means that the base point o' on the end link has a translational velocity along the direction perpendicular to the plane formed by axis of the R_2 and the radius vector ρ_2.

Substitute the above expressions in the form of velocity characteristics into the unit vector set (Eq. 3.4), then VC set of the end link can be obtained as follows:

$$\dot{M}_S = \begin{bmatrix} \dot{t}_1^1(\perp(R_1, \rho_1)) & \dot{t}_2^1(\perp(R_2, \rho_2)) & 0 \\ 0 & 0 & \dot{r}_3^1(\| R_2) \end{bmatrix} \tag{3.5}$$

where, $\dot{M}_S$—VC set of the end link, $\dot{r}^1(\| R_2)$—there exists an angular velocity parallel to axis of the R_2, $\dot{t}^1(\perp(R_1, \rho_1))$—there exists a translational velocity for the base point o′ along the direction perpendicular to the plane formed by axis of R_1 and the radius vector ρ_1, $\dot{t}^1(\perp(R_2, \rho_2))$—there exists a translational velocity for the base point o′ along the direction perpendicular to the plane formed by axis of R_2 and the radius vector ρ_2.

Obviously, the VC set of the end link is precisely the unit vector set of velocity of the end link, which is expressed in the form of velocity characteristic. In other words, the unit vector set could be rewritten as the VC set of the end link.

Unit vector set of the end link velocity is "Union" of unit vector set of each pair velocity (Eq. 3.4) and has the invariance property (excluding singular position). VC set of the end link (Eq. 3.5) is "Union" of VC set of each pair VC set and has the invariance property.

(3) POC set of the end link

Since VC set of the end link has the invariance property (excluding singular position), at nonsingular positions, each element of the VC set corresponds to a finite displacement of the end link:

(a) $\dot{r}^1(\| R_2)$ corresponds to a finite rotation parallel to axis of the R_2 pair. This finite rotation is written as $r^1(\| R_2)$.

(b) $\dot{t}^1(\perp(R_1, \rho_1))$ corresponds to a finite translation of the base point o′ along the direction perpendicular to the plane formed by axis of the R_1 and the radius vector ρ_1. This finite translation can be written as $t^1(\perp(R_1, \rho_1))$.

(c) $\dot{t}^1(\perp(R_2, \rho_2))$ corresponds to a finite translation of the base point o′ along the direction perpendicular to the plane formed by axis of the R_2 and the radius vector ρ_2. This finite translation is written as $t^1(\perp(R_2, \rho_2))$.

Hence, POC set of the planar mechanism's end link (Fig. 3.4) is

$$M_S = \begin{bmatrix} t_1^1(\perp(R_1, \rho_1)) & t_2^1(\perp(R_2, \rho_2)) & t_3^0 \\ r_1^0 & r_2^0 & r_3^1(\| R_2) \end{bmatrix} \tag{3.6}$$

where, M_S—POC set of the end link, $r^1(\| R_2)$—there exists a finite rotation parallel to axis of the R_2, $t^1(\perp(R_1, \rho_1))$—the base point o′ has a finite translation along the direction perpendicular to the plane formed by axis of R_1 and the radius vector ρ_1, $t^1(\perp(R_2, \rho_2))$—the base point o′ has a finite translation along the direction perpendicular to the plane formed by axis of R_2 and the radius vector ρ_2, t_3^0—the end link has no finite translation in corresponding direction, r_1^0,r_2^0—the end link has no rotation in corresponding directions.

It should be noted that there are 3 elements in Eq. (3.6) and DOF of the mechanism is 2. There are only two independent elements in this POC set.

Since the VC set has invariance property (excluding singular position), POC set also has the invariance property. Therefore, the POC set is independent of motion position of the mechanism and the fixed coordinate system.

3.3.2 *VC Set of a Mechanism*

(1) The unit vector set of link velocity

Similar to the Eq. (3.4) derived, the unit vector set of velocity of link i relative to link j can be written as

$$V_M^e = \begin{bmatrix} e_1^t & e_2^t & e_3^t \\ e_1^r & e_2^r & e_3^r \end{bmatrix} \tag{3.7}$$

where, V_M^e—the unit vector set of velocity of link i relative to link j (in short, the unit vector set of link velocity); $e_k^r(k{=}1,2,3)$—the unit vector of the kth angular velocity component of link i; $e_k^t(k{=}1,2,3)$—the unit vector of the kth translational velocity component of base point o′ on link i.

The unit vector set of link velocity depends only on topological structure of a mechanism and is independent of motion position (excluding singular position) of the mechanism (refer to Chaps. 4, 5). Since the topological structure has the invariance (refer to the Chap. 2), this unit vector set also features invariance.

(2) VC set of a mechanism

Since the unit vector set of a link velocity could be rewritten as the VC set described in Sect. 3.3.1, the VC set of the link (such as, the end link of a serial mechanism or the moving platform of a PM) could be defined as [9–13].

$$\dot{M} = \begin{bmatrix} \dot{t}_1^1\,(dir) & \dot{t}_2^1\,(dir) & \dot{t}_3^1\,(dir) \\ \dot{r}_1^1\,(dir) & \dot{r}_2^1\,(dir) & \dot{r}_3^1\,(dir) \end{bmatrix} \tag{3.8}$$

where, $\dot{M}$—the VC set of link i relative to link j, $\dot{r}_k^1(dir)$ $(k = 1, 2, 3)$—the velocity characteristic of the unit vector of the kth angular velocity component of link i, $\dot{t}_k^1(dir)$ $(k = 1, 2, 3)$—the velocity characteristic of the unit vector of the kth translational velocity component of base point o' on link i, (dir)—direction relative to pair axis.

Since mechanism topological structure has the invariance (excluding singular positions), VC set of a mechanism also has the invariance. Hence, the VC set is independent of motion position of the mechanism and the fixed coordinate system (i.e. it is not necessary to establish the fixed coordinate system).

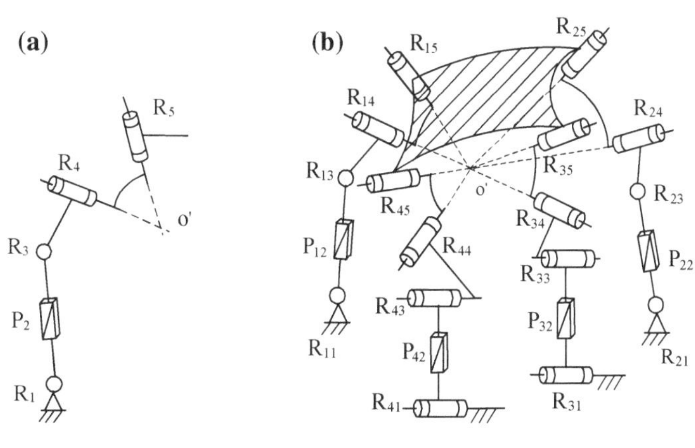

Fig. 3.5 A serial mechanism and a parallel mechanism

3.3.3 POC Set of a Mechanism

Since VC set of the mechanism has the invariance property (excluding singular position), at nonsingular positions, each element of the VC set corresponds to a finite displacement. Therefore, POC set of a general mechanism can be defined as [9–13].

$$M = \begin{bmatrix} t_1^1(dir) & t_2^1(dir) & t_3^1(dir) \\ r_1^1(dir) & r_2^1(dir) & r_3^1(dir) \end{bmatrix} \quad (3.9)$$

where, M—POC set of link i relative to link j, $r_k^1(dir)$ (k = 1, 2, 3)—the kth finite rotation of link i and its direction, $t_k^1(dir)$ (k = 1, 2, 3)—the kth finite translation of point o′ on link i and its direction, (dir)—direction relative to pair axis.

Since VC set has the invariance property (excluding singular position), the corresponding POC set also has the invariance property. The POC set is independent of motion position (excluding singular positions) of the mechanism and the fixed coordinate system (i.e. it is not necessary to establish the fixed coordinate system).

For the serial mechanism ($SOC\{-R(\perp P) \parallel R - \widehat{RR} -\}$) in Fig. 3.5a, point o′ on end link is selected as the base point and POC set of the end link relative to the frame is

$$M_S = \begin{bmatrix} t^2(\perp R_1) \\ r^3 \end{bmatrix}$$

This POC set can be obtained from POC equation for serial mechanisms (refer to Example 4.5 in Chap. 4 for detail). According to this POC set, the end link has three finite rotations. Base point o′ has two finite translations in the plane perpendicular to axis of R_1.

For the parallel mechanism in Fig. 3.5b, point o′ on the moving platform is selected as the base point and POC set of the moving platform relative to the fixed platform is

$$M_{Pa} = \begin{bmatrix} t^1(\perp \diamond (R_{11}, R_{31})) \\ r^3 \end{bmatrix}$$

This POC set can be obtained from POC equation for parallel mechanisms (refer to Example 5.2 in Chap. 5 for detail). According to this POC set, the moving platform has three finite rotations. Base point o′ has one finite translation in the direction perpendicular to axes of R_{11} and R_{31}.

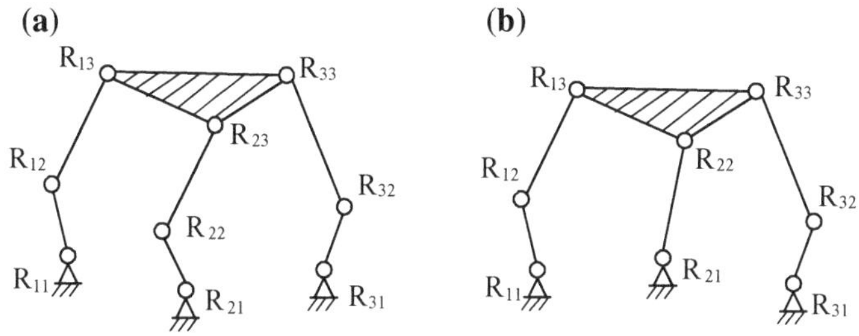

Fig. 3.6 Planar parallel mechanisms

3.4 Dimension of POC Set

The number of independent elements in a POC set is also called dimension of the POC set and is written as $\dim\{M\}$. Dimension of POC set shall not exceed DOF of the mechanism [9–13], i.e.

$$\dim\{M\} = \dim\{M(t)\} + \dim\{M(r)\} \leq DOF \tag{3.10}$$

where, $\dim\{M\}$- dimension of the POC set; $\dim\{M(t)\}$- number of independent translational elements in the POC set ($\dim\{M(t)\} = 0, 1, 2$ or 3), $\dim\{M(r)\}$- number of independent rotational elements in the POC set ($\dim\{M(r)\}$ = 0, 1, 2 or 3).

It is specified that a non-independent element of the POC set is put into a brace.

For the planar 8-bar linkage in Fig. 3.6a, POC set of its moving platform is (the calculation process is omitted here)

$$M_{Pa} = \begin{bmatrix} t^1(\perp R_{11}) & t^1(\perp R_{12}) \\ \multicolumn{2}{c}{r^1(\parallel R_{13})} \end{bmatrix}$$

where, $t^1(\perp R)$—abbreviated form of $t^1(\perp(R, \rho))$ (It means that there exists a finite translation in a plane perpendicular to axis of R pair).

Since DOF of the mechanism is three, POC set has three independent elements. There is $\dim\{M_{Pa}\} = 3 =$ DOF.

For the planar 7-bar linkage in Fig. 3.6b, DOF is two and POC set of its moving platform is (the calculation process is omitted here)

$$M_{Pa} = \begin{bmatrix} t^1(\perp R_{11}) & \{t^1(\perp R_{12})\} \\ \multicolumn{2}{c}{r^1(\parallel R_{12})} \end{bmatrix}$$

Since DOF of the mechanism is two, the POC set has two independent elements and one non-independent element. In this case, $t^1(\perp R_{11})$ and $r^1(\parallel R_{12})$ are selected as independent elements and $\{t^1(\perp R_{12})\}$ is a non-independent element. There is $\dim\{M_{Pa}\} = 2 =$ DOF.

Table 3.2 Basic types of POC set [9–13]

dim{M}	1	2	3	4	5	6
M	$\begin{bmatrix} t^1(dir) \\ r^0 \end{bmatrix}$	$\begin{bmatrix} t^2(dir) \\ r^0 \end{bmatrix}$	$\begin{bmatrix} t^3 \\ r^0 \end{bmatrix}$	$\begin{bmatrix} t^3 \\ r^1(dir) \end{bmatrix}$	$\begin{bmatrix} t^3 \\ r^2(dir) \end{bmatrix}$	$\begin{bmatrix} t^3 \\ r^3 \end{bmatrix}$
	$\begin{bmatrix} t^0 \\ r^1(dir) \end{bmatrix}$	$\begin{bmatrix} t^1(dir) \\ r^1(dir) \end{bmatrix}$	$\begin{bmatrix} t^2(dir) \\ r^1(dir) \end{bmatrix}$	$\begin{bmatrix} t^2(dir) \\ r^2(dir) \end{bmatrix}$	$\begin{bmatrix} t^2(dir) \\ r^3 \end{bmatrix}$	
		$\begin{bmatrix} t^0 \\ r^2(dir) \end{bmatrix}$	$\begin{bmatrix} t^1(dir) \\ r^2(dir) \end{bmatrix}$	$\begin{bmatrix} t^1(dir) \\ r^3 \end{bmatrix}$		
			$\begin{bmatrix} t^0 \\ r^3 \end{bmatrix}$			

Notes t^0—no such finite translation, r^0—no such finite rotation

3.5 Basic types of POC Set

If a POC set contains only independent elements, it is called a basic type of POC set. There are 15 basics types of POC set in total, as shown Table 3.2 [9–13].

3.6 Summary

(1) Based on the velocity analysis of kinematic pairs and the unit vector of velocity expressed in the form of velocity characteristics, VC sets of kinematic pairs are defined. Subsequently, POC sets (Eqs. 3.1–3.3) of kinematic pairs are defined based on invariance property of the topological structure. POC set of R pair/H pair has multiplicity (Table 3.1).

(2) Based on the velocity analysis of mechanism and the unit vector of velocity expressed in the form of velocity characteristics, VC set (Eq. 3.8) of mechanism is defined. Subsequently, POC set (Eq. 3.9) of mechanism is defined based on invariance property of the mechanism topological structure.

(3) Since unit vector set of velocity is related only to mechanism topological structure, both VC set and POC set of a mechanism have invariance property. They are independent of motion position (excluding singular position) of the mechanism and the fixed coordinate system (i.e. it is not necessary to establish the fixed coordinate system). Therefore, POC set is a geometric tool to describe mechanism motion characteristics.

(4) Dimension of the POC set is not greater than DOF of the mechanism (Eq. 3.10). POC set may contain non-independent element(s). There are only 15 basic types of POC set containing only independent elements (Table 3.2).

(5) Unit vector of velocity in the form of velocity characteristics reveals the internal relation between mechanism topological structure and motion characteristics and provides a theoretical basis for operations of POC sets based on topological structure.

References

1. Hunt KH (1987) Kinematic geometry of mechanisms. Clarendon Press, Oxford
2. Huang Z, Li QC (2002) General methodology for type synthesis of symmetrical lower-mobility parallel manipulators and several novel manipulators. Int J Robot Res 21:131–145
3. Kong X, Gosselin C (2007) Type synthesis of parallel mechanisms. Springer, Heidelberg
4. Meng X, Gao F, Shengfu WuS, Ge QJ (2014) Type synthesis of parallel robotic mechanisms: framework and brief review. Mech Mach Theor 78:177–186
5. Herve JM (1999) The lie group of rigid body displacements, a fundamental tool for mechanism design. Mech Mach Theor 34:719–730
6. Selig J (1996) Geometrical methods in robotics. Springer, New York
7. Meng J, Liu GF, Li ZX (2007) A geometric theory for analysis and synthesis of sub-6 dof parallel manipulators. IEEE Trans Rob 23:625–649
8. Gogu G (2007) Structural synthesis of parallel robots: part 1: methodology. Springer, Dordrecht
9. Yang T-L, Jin Q et al (2001) A general method for type synthesis of rank-deficient parallel robot mechanisms based on SOC unit. Mach Sci Technol 20(3):321–325
10. Yang T-L, Jin Q, Liu A-X, Shen H-P, Luo Y-F (2002) Structural synthesis and classification of the 3-DOF translation parallel robot mechanisms based on the unites of single-open chain. Chinese J Mech Eng 38(8):31–36. doi:10.3901/JME.2002.08.031
11. Jin Q, Yang T-L (2004) Theory for topology synthesis of parallel manipulators and its application to three-dimension-translation parallel manipulators. ASME J Mech Des 126:625–639
12. Yang T-L (2004) Theory of topological structure for robot mechanisms. China machine press, Beijing
13. Yang T-L, Liu A-X, Luo Y-F, Shen H-P et al (2009) Position and orientation characteristic equation for topological design of robot mechanisms. ASME J Mech Des 131:021001-1–021001-17
14. Yang T-L, Sun D-J (2012) A general DOF formula for parallel mechanisms and multi-loop spatial mechanisms. ASME J Mech Robot 4(1):011001-1–011001-17
15. Yang T-L, Liu A-X, Shen H-P, Luo Y-F et al (2013) On the correctness and strictness of the position and orientation characteristic equation for topological design of robot mechanisms. ASME J Mech Robot 5:021009-1–021009-18
16. Yang T-L, Liu A-X, Shen H-P et al (2015) Composition principle based on SOC unit and coupling degree of BKC for general spatial mechanisms. The 14th IFToMM World Congress, Taipei, Taiwan, 25–30 Oct 2015. doi: 10.6567/IFToMM.14TH.WC.OS13.135
17. Yang T-L, Liu A-X, Shen H-P et al (2012) Theory and application of robot mechanism topology. China Science press, Beijing

Chapter 4
Position and Orientation Characteristics Equation for Serial Mechanisms

4.1 Introduction

POC equation for serial mechanisms and the corresponding operation rules are established in this chapter. The POC equation for serial mechanisms reveals the mapping relations among topological structure, POC set and DOF of serial mechanisms.

For a serial mechanism whose topological structure is known, there are the following four types of methods to determine the motion characteristics of the end link relative to the frame link.

(1) Method based on screw theory

Motion screw system of the end link can be obtained through linear operation of motion screw systems. This method depends on motion position of the mechanism and the fixed coordinate system [1–4].

(2) Method based on displacement subgroup/submanifold

For serial mechanisms fulfilling algebra structures of Lie group, the motion characteristics of the end link can be obtained through "Union" operation rules of displacement subgroups using synthetic arguments or tables of compositions of subgroups, which is independent of motion position of a mechanism [4–8]. For serial mechanisms not fulfilling the algebra structures of Lie group, its motion characteristics can be obtained by the method based on displacement submanifold. But this method depends on motion position of a mechanism [8].

(3) Method based on linear transformation

The velocity space of the end link relative to the frame link can be obtained through linear operation. This method depends motion position of the mechanism and fixed coordinate system [4, 9].

T.-L. Yang et al., *Topology Design of Robot Mechanisms*,
https://doi.org/10.1007/978-981-10-5532-4_4

(4) Method based on POC equation for serial mechanisms

POC set of the end link can be determined through forward operation of POC equation (Eq. 4.4) and its 10 “Union” operation rules for serial mechanisms [10–19]. This method is a geometrical method. It is independent of motion position of the mechanism and the fixed coordinate system (i.e. it is not necessary to establish the fixed coordinate system).

This chapter focuses on establishment of POC equation for serial mechanisms and its operation rules. Firstly, VC equation for serial mechanisms and its operation rules are established based on (a) velocity analysis, (b) invariance property of mechanism topological structure and (c) unit vector set of end link velocity being “union” of unit vector set of each pair’s velocity. Then based on invariance property of VC set and one-to-one correspondence between elements of VC set and elements of POC set, POC equation for serial mechanisms and its operation rules are obtained.

4.2 Velocity Analysis of Serial Mechanisms

According to principle of motion synthesis, velocity of the end link of the serial mechanism in Fig. 4.1 can be written as:

$$\begin{cases} \omega = \sum\limits_{i=1}^{n} \left|\omega_{i,i-1}\right| e_{R_i} \\ v_{o'} = \sum\limits_{i=1}^{n} \left|v_{i,i-1}\right| e_{P_i} + \sum\limits_{i=1}^{n} \left|\omega_{i,i-1}\right| \cdot \left|\rho_{d_i-o'}\right| \left(e_{R_i} \times e_{\rho_i}\right) \end{cases} \tag{4.1}$$

where, ω—angular velocity of end link, $\omega_{i,i-1}$—relative angular velocity between the two links connected by the ith pair, e_{R_i}—unit vector of axis of the ith R pair, n—number of links, $v_{o'}$—translational velocity of the base point o′ on end link, $v_{i,i-1}$—relative translational velocity between the two links connected by the ith P pair, e_{P_i}—unit vector of axis of the ith P pair, $\rho_{d_i-o'}$—radius vector from point d_i on axis of the ith R pair to base point o′, e_{ρ_i}—unit vector of the ith radius vector.

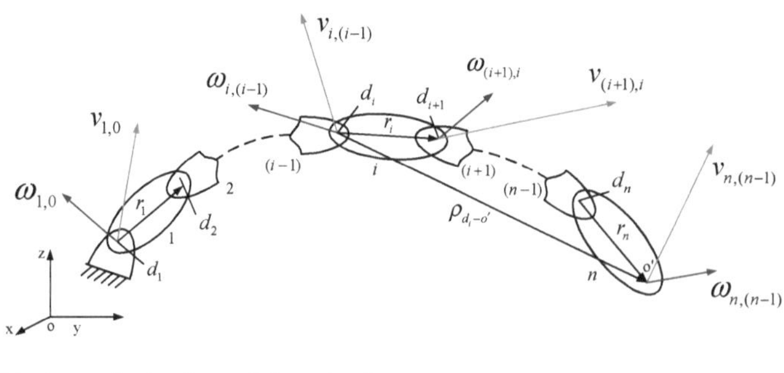

Fig. 4.1 Velocity analysis of a serial mechanism

According to Eq. (4.1),

(1) Angular velocity ω of the end link depends on relative angular velocity $\omega_{i,i-1}$ of each pair.

(2) Translational velocity $v_{o'}$ of base point o′ on the end link depends not only on relative translational velocity $v_{i,i-1}$ of P pair, but also on translational velocity $\left(\sum_{i=1}^{n} \omega_{i,i-1} \times r_{d_i-o'}\right)$ derived from relative angular velocity of R (or H) pair.

(3) There are two types of relative translational velocity along pair axis: inherent relative translational velocity along axis of P pairs and associated translational velocity $\left(\sum_{k=1}^{i} h\omega_{k,k-1}\right.$, h is thread pitch of the H pair$\left.\right)$ along axis of H pairs.

4.3 VC Equation for Serial Mechanisms and Its Operation Rules

4.3.1 VC Equation for Serial Mechanisms

According to Eq. (4.1),

(1) The velocity of the end link of an serial mechanism is the linear combination of relative velocities of each kinematic pair, which depends on motion position of the mechanism and the fixed coordinate system.

(2) The unit vector set of the end link velocity of an serial mechanism is "union" of the unit vector sets of each pair velocity and could be written as

$$V_S^e = \bigcup_{i=1}^{m} V_{J_i}^e \tag{4.2}$$

where, V_S^e—unit vector set of the end link velocity (refer to Eq. (3.7)); $V_{J_i}^e$—unit vector set of the *i*th kinematic pair for the same base point o′; m—number of kinematic pairs.

Equation (4.2) depends only on topological structure. Due to the topological structure invariance (excluding singular positions) of a mechanism described in Chap. 2, the unit vector set of the end link velocity has the invariance and is independent of motion position of a mechanism.

(3) Since the unit vector set of velocity could be rewritten as the VC set described in chap. 3, the VC set of the end link of the serial mechanism is "union" of the VC sets of each pair, which also has the invariance and is independent of motion position of a mechanism (excluding singular positions) and the fixed coordinate system.

Therefore, VC set of the end link relative to the frame link is "union" of each pair's VC set [10–14], i.e.

$$\dot{M}_S = \bigcup_{i=1}^{m} \dot{M}_{J_i} = \bigcup \dot{M}_{sub-SOC_j} \tag{4.3}$$

where, $\dot{M}_S$—VC set of the end link relative to the framce link, $\dot{M}_{J_i}$—VC set of the ith kinematic pair, $\dot{M}_{sub-SOC_j}$—VC set of the jth sub-SOC_j, m—number of kinematic pairs.

Equation (4.3) is called VC equation for serial mechanisms. It should be noted that:

(1) VC set of each kinematic pair is related to the same base point o′ on the end link (i.e. origin of the moving coordinate system on the end link).

(2) Number of independent elements in the end link's VC set should not exceed DOF of the mechanism (Eq. (3.10) in Chap. 3).

4.3.2 Operation Rules for VC Equation

For VC equation for serial mechanisms (Eq. 4.3), there are the following three types of symbolic operation rules [10–14]:

(1) Operation rules for "union" of rotational elements [linear operation: Eqs. (4.3a)–(4.3d)].

(2) Operation rules for "union" of translational elements [linear operation: Eqs. (4.3e)–(4.3h)].

(3) Rules for selection of independent elements [nonlinear criteria: Eqs. (4.3i), (4.3j)].

A. **Operation rules for "union" of rotational elements**

(1) $[\dot{r}^1] \bigcup [\dot{r}^1]$

$$[\dot{r}^1(\| R_i)] \bigcup [\dot{r}^1(\| R_{i+1})] = \begin{cases} [\dot{r}^1(\| R_i)], & if\, R_i \| R_{i+1} (or\, R_i | R_{i+1}). \\ [\dot{r}^2(\| \diamondsuit(R_i, R_{i+1}))], & if\, R_i \nparallel R_{i+1}. \end{cases} \tag{4.3a}$$

where, $\dot{r}^1(\| R_i)$—one rotational element (unit vector of angular velocity) in a direction parallel to axis of R_i, $\dot{r}^2(\| \diamondsuit(R_i, R_{i+1}))$—two rotational elements (unit vector of angular velocity) in any direction within a plane parallel to axis of R_i and axis of R_{i+1}.

(2) $[\dot{r}^1] \bigcup [\dot{r}^2]$

$$[\dot{r}^1(\| R_i)] \bigcup [\dot{r}^2(\| \diamondsuit(R_{i+1}, R_{i+2}))] = \begin{cases} [\dot{r}^2(\| \diamondsuit(R_{i+1}, R_{i+2}))], & if\, R_i \| \diamondsuit(R_{i+1}, R_{i+2}). \\ [\dot{r}^3], & if\, R_i \nparallel \diamondsuit(R_{i+1}, R_{i+2}). \end{cases} \tag{4.3b}$$

where, $\dot{r}^3$—three rotational elements (unit vector of angular velocity) in any direction within a 3D space.

(3) $[\dot{r}^2]\bigcup[\dot{r}^2]$

$$\begin{aligned}&[\dot{r}^2(\parallel\Diamond(R_{i-1},R_i))]\bigcup[\dot{r}^2(\parallel\Diamond(R_{i+1},R_{i+2}))]\\&=\begin{cases}[\dot{r}^2(\parallel\Diamond(R_{i+1},R_{i+2}))], & \text{if } \Diamond(R_{i-1},R_i)\parallel\Diamond(R_{i+1},R_{i+2}).\\ [\dot{r}^3], & \text{if} \Diamond(R_{i-1},R_i)\nparallel\Diamond(R_{i+1},R_{i+2}).\end{cases}\end{aligned} \tag{4.3c}$$

(4) $[\dot{r}^3]\bigcup[\dot{r}^k]$

$$[\dot{r}^3]\bigcup[\dot{r}^k]=[\dot{r}^3],\quad \mathrm{k}=0,1,2,3 \tag{4.3d}$$

B. **Operation rules for "union" of translational elements**

(1) $[\dot{t}^1]\bigcup[\dot{t}^1]$

$$[\dot{t}^1(\parallel P_i^*)]\bigcup\left[\dot{t}^1(\parallel P_j^*\right]=\begin{cases}[\dot{t}^1(\parallel P_i^*)], & \text{if } P_i^*|P_j^*.\\ \left[\dot{t}^2(\parallel\Diamond(P_i^*,P_j^*))\right], & \text{if } P_i^*\nparallel P_j^*.\end{cases} \tag{4.3e}$$

where, P^*—translation of a P pair or derivative translation of an R (or H) pair, $\dot{t}^1(\parallel P_i^*)$—one translational element (unit vector of translational velocity) in a direction parallel to P_i^*, $\dot{t}^2(\parallel\Diamond(P_i^*,P_j^*))$—two translational elements (unit vector of translational velocity) in any direction within a plane parallel to P_i^* and P_j^*.

(2) $[\dot{t}^1]\bigcup[\dot{t}^2]$

$$[\dot{t}^1(\parallel P_i^*)]\bigcup[\dot{t}^2(\parallel\Diamond(P_{i+1}^*,P_{i+2}^*))]=\begin{cases}[\dot{t}^2(\parallel\Diamond(P_{i+1}^*,P_{i+2}^*))], & \text{if } P_i^*\parallel\Diamond(P_{i+1}^*,P_{i+2}^*).\\ [\dot{t}^3], & \text{if } P_i^*\nparallel\Diamond(P_{i+1}^*,P_{i+1}^*).\end{cases} \tag{4.3f}$$

where, $\dot{t}^3$— three translational elements (unit vector of translational velocity) in any direction within a 3D space.

(3) $[\dot{t}^2]\bigcup[\dot{t}^2]$

$$\begin{aligned}&[\dot{t}^2(\parallel\Diamond(P_{i-1}^*,P_i^*))]\bigcup[\dot{t}^2(\parallel\Diamond(P_{i+1}^*,P_{i+2}^*))]\\&=\begin{cases}[\dot{t}^2(\parallel\Diamond(P_{i+1}^*,P_{i+2}^*))], & \text{if } \Diamond(P_{i-1}^*,P_i^*)\parallel\Diamond(P_{i+1}^*,P_{i+2}^*).\\ [\dot{t}^3], & \text{if } \Diamond(P_{i-1}^*,P_i^*)\nparallel\Diamond(P_{i+1}^*,P_{i+2}^*).\end{cases}\end{aligned} \tag{4.3g}$$

(4) $[\dot{t}^3]\bigcup[\dot{t}^k]$

$$[\dot{t}^3]\bigcup[\dot{t}^k] = [\dot{t}^3], \quad \mathrm{k} = 0, 1, 2, 3. \tag{4.3h}$$

C. **Rules for selection of independent rules (nonlinear criteria)**

(1) Number of independent elements in VC set should not exceed DOF of the mechanism. (4.3i)

(2) In order to reduce the number of non-independent elements, selection of independent element shall abide by the following criteria:

(a) The rotation element without derivative translation shall be preferably selected.

(b) Translation element of P pair shall be preferably selected. (4.3j)

It should be noted that these ten operation rules above depend on only topological structure and are independent of motion position of the mechanism and it is not necessary to establish the fixed coordinate system.

4.4 POC Equation for Serial Mechanisms and Its Operation Rules

4.4.1 POC Equation for Serial Mechanisms

Due to the VC set has invariance (excluding singular positions) described in Sect. 4.3.1, each element of the VC set of Eq. (4.3) corresponds to a finite displacement. Correspondingly, the POC equation for serial mechanisms can be written as follows [10–14]

$$M_S = \bigcup_{i=1}^{m} M_{J_i} = \bigcup M_{sub-SOC_j} \tag{4.4}$$

where, M_S—POC set of the end link relative to the frame link (also called POC set of the serial mechanism), M_{J_i}—POC set of the ith kinematic pair (Eqs. (3.1)–(3.3) in Chap. 3), $M_{sub-SOC_j}$—POC set of the jth sub-SOC (as shown in Table 4.1).

Table 4.1 POC sets of sub-SOCs [12–14]

Sub-SOCs	$SOC\{-R \parallel R-\}$ $SOC\{-R\perp P-\}$	$SOC\{-R \parallel R \parallel R-\}$, $SOC\{-R \parallel R\perp P-\}$, $SOC\{-P\perp R\perp P-\}$.	$SOC\{\diamondsuit(P, P, \cdots, P)\}$
POC set	No. 1	No. 2	No. 3
	$\begin{bmatrix} t^2(\perp R) \\ r^1(\parallel R) \end{bmatrix}$	$\begin{bmatrix} t^2(\perp R) \\ r^1(\parallel R) \end{bmatrix}$	$\begin{bmatrix} t^2 \\ r^0 \end{bmatrix}$
	$\text{Dim}\{\text{M}_\text{S}\} = 2$	$\text{Dim}\{\text{M}_\text{S}\} = 3$	$\text{Dim}\{\text{M}_\text{S}\} = 2$
	No. 1[a]	No. 2[a]	No. 3[a]
	$\begin{bmatrix} t^1(\perp R) \\ r^1(\parallel R) \end{bmatrix}$	$\begin{bmatrix} t^2(\perp R) \\ r^1(\parallel R) \end{bmatrix}$	$\begin{bmatrix} t^2 \\ r^0 \end{bmatrix}$
	$\text{Dim}\{\text{M}_\text{S}\} = 2$	$\text{Dim}\{\text{M}_\text{S}\} = 3$	$\text{Dim}\{\text{M}_\text{S}\} = 2$
Sub-SOCs	$SOC\{-R \parallel P-\}$	$SOC\{-\overbrace{RR}-\}$	$SOC\{-\overbrace{RRR}-\}$
POC set	No. 4	No. 5	No. 6
	$\begin{bmatrix} t^1(\parallel P)\ t^1(\perp R) \\ r^1(\parallel R) \end{bmatrix}$	$\begin{bmatrix} t^2(\perp \rho) \\ r^2 \end{bmatrix}$	$\begin{bmatrix} t^2(\perp \rho) \\ r^3 \end{bmatrix}$
	$\text{Dim}\{\text{M}_\text{S}\} = 2$	$\text{Dim}\{\text{M}_\text{S}\} = 2$	$\text{Dim}\{\text{M}_\text{S}\} = 3$
	No. 4[a]	No. 5[a]	No. 6[a]
	$\begin{bmatrix} t^1(\parallel P) \\ r^1(\parallel R) \end{bmatrix}$	$\begin{bmatrix} t^0 \\ r^2 \end{bmatrix}$	$\begin{bmatrix} t^0 \\ r^3 \end{bmatrix}$
	$\text{Dim}\{\text{M}_\text{S}\} = 2$		$\text{Dim}\{\text{M}_\text{S}\} = 3$
Sub-SOCs	$SOC\{-R \vert H-\}$	$SOC\{-H \parallel H-\}$	$SOC\{-H \parallel H \parallel H \parallel H-\}$
POC set	No. 7	No. 8	No. 9
	$\begin{bmatrix} t^1(\parallel H)\ t^1(\perp R) \\ r^1(\parallel R) \end{bmatrix}$	$\begin{bmatrix} t^1(\parallel H)\ t^1(\perp H) \\ r^1(\parallel H) \end{bmatrix}$	$\begin{bmatrix} t^3 \\ r^1(\parallel H) \end{bmatrix}$
	$\text{Dim}\{\text{M}_\text{S}\} = 2$		$\text{Dim}\{\text{M}_\text{S}\} = 4$
	No. 7[a]	No. 8[a]	No. 9[a]
	$\begin{bmatrix} t^1(\parallel H) \\ r^1(\parallel R) \end{bmatrix}$	$\begin{bmatrix} t^1(\parallel H) \\ r^1(\parallel H) \end{bmatrix}$	$\begin{bmatrix} t^3 \\ r^1(\parallel H) \end{bmatrix}$
	$\text{Dim}\{\text{M}_\text{S}\} = 2$		$\text{Dim}\{\text{M}_\text{S}\} = 4$

[a]base point o′ is on axis of the end pair

It should be noted that:

(1) POC set of each kinematic pair is related to the same base point o′ on the end link (i.e. origin of the moving coordinate system).

(2) Number of independent elements in the end link's POC set should not exceed DOF of the mechanism (Eq. (3.10) in Chap. 3).

Equation (4.4) is called the POC equation for serial mechanisms, which also has the invariance and is independent of motion position of a mechanism (excluding singular positions) and the fixed coordinate system. It reveals the function relations among topological structure, POC set and DOF of mechanisms. Forward operation of this POC equation can be used in mechanism structure analysis, i.e. determining

POC set and DOF of a serial mechanism when its topological structure is known (refer to Sect. 4.5). Inverse operation of this POC equation can be used in mechanism structure synthesis, i.e. determining topological structure of a serial mechanism when its POC set and DOF are known (refer to Chap. 8).

4.4.2 Operation Rules for POC Equation

Based one-to-one correspondence between POC set and VC set of the end link, the following three types of symbolic operation rules for POC equation can be derived from operation rules for VC equation (Eq. 4.3) [12–14]:

(1) Operation rules for "union" of rotational elements [linear operation: Eqs. (4.4a)–(4.4d)].

(2) Operation rules for "union" of translational elements [linear operation: Eqs. (4.4e)–(4.4h)].

(3) Rules for selection of independent elements [nonlinear criteria: Eqs. (4.4i), (4.4j)].

A. **Operation rules for "union" of rotational elements**

(1) $[r^1]\bigcup[r^1]$

$$[r^1(\| R_i)]\bigcup[r^1(\| R_{i+1})] = \begin{cases} [r^1(\| R_i)], & \textit{if } R_i \| R_{i+1} (\textit{or } R_i | R_{i+1}). \\ [r^2(\| \diamondsuit(R_i, R_{i+1}))], & \textit{if } R_i \nparallel R_{i+1}. \end{cases} \tag{4.4a}$$

where, $r^1(\| R_i)$—the end link has a finite rotation in a direction parallel to axis of R_i, $r^2(\| \diamondsuit(R_i, R_{i+1}))$—the end link has two finite rotations in directions within a plane parallel to axis of R_i and axis of R_{i+1}.

(2) $[r^1]\bigcup[r^2]$

$$[r^1(\| R_i)]\bigcup[r^2(\| \diamondsuit(R_{i+1}, R_{i+2}))] = \begin{cases} [r^2(\| \diamondsuit(R_{i+1}, R_{i+2}))], & \textit{if } R_i \| \diamondsuit(R_{i+1}, R_{i+2}). \\ [r^3], & \textit{if } R_i \nparallel \diamondsuit(R_{i+1}, R_{i+2}). \end{cases} \tag{4.4b}$$

where, r^3—the end link has three finite rotation.

(3) $[r^2]\bigcup[r^2]$

$$\begin{aligned}&[r^2(\parallel \diamondsuit(R_{i-1},R_i))]\bigcup[r^2(\parallel \diamondsuit(R_{i+1},R_{i+2}))]\\&=\begin{cases}[r^2(\parallel \diamondsuit(R_{i+1},R_{i+2}))], & \text{if } \diamondsuit(R_{i-1},R_i)\parallel\diamondsuit(R_{i+1},R_{i+2}).\\ [r^3], & \text{if } \diamondsuit(R_{i-1},R_i)\nparallel\diamondsuit(R_{i+1},R_{i+2}).\end{cases}\end{aligned}\tag{4.4c}$$

(4) $[r^3]\bigcup[r^k]$

$$[r^3]\bigcup[r^k]=[r^3],\quad \mathrm{k}=0,1,2,3.\tag{4.4d}$$

B. **Operation rules for "union" of translational elements**

(1) $[t^1]\bigcup[t^1]$

$$[t^1(\parallel P_i^*)]\bigcup\left[t^1(\parallel P_j^*\right]=\begin{cases}[t^1(\parallel P_i^*)], & \text{if } P_i^*|P_j^*.\\ \left[t^2(\parallel \diamondsuit(P_i^*,P_j^*))\right], & \text{if } P_i^*\nparallel P_j^*.\end{cases}\tag{4.4e}$$

where, P^*—translation of a P pair or derivative translation of an R (or H) pair, $t^1(\parallel P_i^*)$—the end link has a finite translation in a direction parallel to P_i^*, $t^2(\parallel \diamondsuit(P_i^*,P_j^*))$—the end link has two finite translations in directions within a plane parallel to P_i^* and P_j^*.

(2) $[t^1]\bigcup[t^2]$

$$[t^1(\parallel P_i^*)]\bigcup[t^2(\parallel \diamondsuit(P_{i+1}^*,P_{i+2}^*))]=\begin{cases}[t^2(\parallel \diamondsuit(P_{i+1}^*,P_{i+2}^*))], & \text{if } P_i^*\parallel\diamondsuit(P_{i+1}^*,P_{i+2}^*).\\ [t^3], & \text{if } P_i^*\nparallel\diamondsuit(P_{i+1}^*,P_{i+1}^*).\end{cases}\tag{4.4f}$$

where, t^3—the end link has three finite translations.

(3) $[t^2]\bigcup[t^2]$

$$\begin{aligned}&[t^2(\parallel \diamondsuit(P_{i-1}^*,P_i^*))]\bigcup[t^2(\parallel \diamondsuit(P_{i+1}^*,P_{i+2}^*))]\\&=\begin{cases}[t^2(\parallel \diamondsuit(P_{i+1}^*,P_{i+2}^*))], & \text{if } \diamondsuit(P_{i-1}^*,P_i^*)\parallel\diamondsuit(P_{i+1}^*,P_{i+2}^*).\\ [t^3], & \text{if } \diamondsuit(P_{i-1}^*,P_i^*)\nparallel\diamondsuit(P_{i+1}^*,P_{i+2}^*).\end{cases}\end{aligned}\tag{4.4g}$$

(4) $[t^3]\bigcup[t^k]$

$$[t^3]\bigcup[t^k]=[t^3],\quad \mathrm{k}=0,1,2,3.\tag{4.4h}$$

C. **Rules for selection of independent rules** (**nonlinear criteria**)

(1) Number of independent elements in POC set should not exceed DOF of the mechanism. (4.4i)

(2) In order to reduce the number of non-independent elements, selection of independent element shall abide by the following criteria:

   (a) The rotation element without derivative translation shall be preferably selected.

   (b) Translation element of P pair shall be preferably selected. (4.4j)

It should be noted that these ten operation rules above depend on only topological structure and are independent of motion position of the mechanism and it is not necessary to establish the fixed coordinate system.

4.5 Calculation of POC Set for Mechanisms

4.5.1 Main Steps

(1) Determine topological structure of the serial mechanism.

(2) Select base point o′ on end link (origin of the moving coordinate system). Generally, point o′ is on axis of the end pair or at the intersection point of several pair axes.

(3) Substitute POC set of each pair or sub-SOC in Table 4.1 (from frame link to end link) into Eq. (4.4).

(4) Determine rotational elements of the end link's POC set based on Eqs. (4.4a)–(4.4d).

(5) Determine translational elements of the end link's POC set based on Eqs. (4.4e)–(4.4h).

(6) Select independent elements of the end link's POC set based on Eqs. (4.4i) and (4.4j). Put non-independent element in a brace.

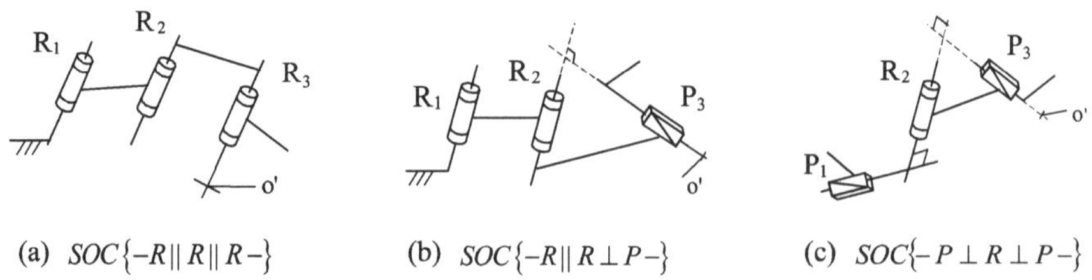

Fig. 4.2 $SOC\{-R \parallel R \parallel R-\}$ and the equivalent SOCs

4.5.2 Examples

Example 4.1 Determine POC set of the serial mechanism in Fig. 4.2a.

(1) Determine topological structure of the mechanism:

$$SOC\{-R_1 \parallel R_2 \parallel R_3-\}.$$

(2) Select point o′ on axis of the end pair (R_3) as base point, as shown in Fig. 4.2a.

(3) Establish POC equation.

Substitute POC set of each R pair (Eq. 3.2) into Eq. (4.4), there is

$$\begin{aligned} M_S &= \begin{bmatrix} t^1(\perp(R_1,\rho_1)) \\ r^1(\parallel R_1) \end{bmatrix} \bigcup \begin{bmatrix} t^1(\perp(R_2,\rho_2)) \\ r^1(\perp R_2) \end{bmatrix} \bigcup \begin{bmatrix} t^0 \\ r(\parallel R_3) \end{bmatrix} \\ &= \begin{bmatrix} t^1(\perp(R_1,\rho_1)) \bigcup t^1(\perp(R_2,\rho_2)) \bigcup \quad t^0 \\ r^1(\parallel R_1) \quad \bigcup \quad r^1(\parallel R_2) \quad \bigcup r^1(\parallel R_3) \end{bmatrix} \end{aligned}$$

(4) Determine rotational elements of the POC set.

Based on the topological structure $(R_1 \parallel R_2 \parallel R_3)$, and the operation rules Eqs. (4.4a) and (4.4j), the above equation can be rewritten as

$$\begin{aligned} M_S &= \begin{bmatrix} t^1(\perp(R_1,\rho_1)) \bigcup t^1(\perp(R_2,\rho_2)) \bigcup \quad t^0 \\ r^1(\parallel R_1) \quad \bigcup \quad r^1(\parallel R_2) \quad \bigcup r^1(\parallel R_3) \end{bmatrix} \\ &= \begin{bmatrix} t^1(\perp(R_1,\rho_1)) \bigcup t^1(\perp(R_2,\rho_2)) \bigcup t^0 \\ r^1(\parallel R_3) \end{bmatrix} \end{aligned}$$

where, $r^1(\parallel R_3)$—the end link has one finite rotation in a direction parallel to axis of R_3.

(5) Determine translational elements of the POC set.

Based on the topological structure $(R_1 \parallel R_2 \parallel R_3)$, and the operation rules Eqs. (4.4e) and (4.4j), the above equation can be further rewritten as

$$M_S = \begin{bmatrix} t^1(\perp(R_1,\rho_1)) \bigcup t^1(\perp(R_2,\rho_2)) \bigcup t^0 \\ r^1(\parallel R_3) \end{bmatrix} = \begin{bmatrix} t^2(\perp R_1) \\ r^1(\parallel R_3) \end{bmatrix}$$

where, $t^2(\perp R_1)$—the end link has two finite translations in directions within a plane perpendicular to axis of R_1.

(6) Select independent elements of the POC set

DOF of the mechanism is three, so there may be up to three independent elements in the POC set.

Therefore, POC set of the end link is

$$M_S = \begin{bmatrix} t^2(\perp R_1) \\ r^1(\parallel R_3) \end{bmatrix}$$

Since DOF = 3, there may be three independent element in the POC set, i.e. $\dim.\{M_S\} = 3 = DOF$.

Similarly, it can be proved that the serial mechanism in Fig. 4.2b and the mechanism in Fig. 4.2c have the same POC set as the mechanism in Fig. 4.2a. They are called topologically equivalent mechanisms.

Example 4.2 Determine POC set of the serial mechanism in Fig. 4.3.

(1) Determine topological structure of the mechanism:

$$SOC\{-\overbrace{R_1R_2R_3}-\}.$$

(2) Select point o′ outside the axis of the end pair (R_3) as base point, as shown in Fig. 4.3.

(3) Establish POC equation.

Substitute POC set of each R pair (Eq. 3.2) into Eq. (4.4), there is

$$M_S = \begin{bmatrix} t^1(\perp(R_1,\rho)) \\ r^1(\parallel R_1) \end{bmatrix} \bigcup \begin{bmatrix} t^1(\perp(R_2,\rho)) \\ r^1(\perp R_2) \end{bmatrix} \bigcup \begin{bmatrix} t^1(\perp(R_3,\rho)) \\ r(\parallel R_3) \end{bmatrix}$$
$$= \begin{bmatrix} t^1(\perp(R_1,\rho)) \bigcup t^1(\perp(R_2,\rho)) \bigcup t^1(\perp(R_3,\rho)) \\ r^1(\parallel R_1) \quad \bigcup \quad r^1(\parallel R_2) \quad \bigcup \quad r^1(\parallel R_3) \end{bmatrix}$$

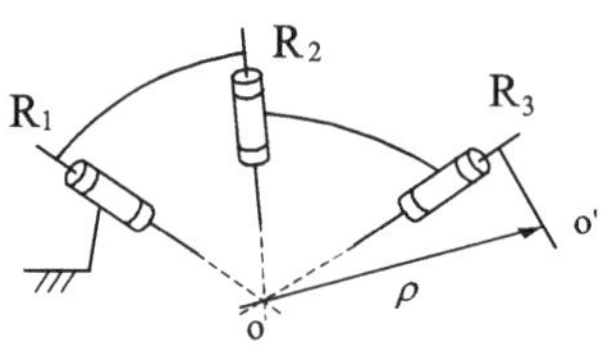

Fig. 4.3 $SOC\{-\overbrace{RRR}-\}$

(4) Determine rotational elements of the POC set.

Based on the topological structure $(R_1 /\!\!/ R_2 /\!\!/ R_3)$, and the operation rules Eqs. (4.4a) and (4.4b), the above equation can be rewritten as

$$M_S = \begin{bmatrix} t^1(\perp(R_1,\rho)) \bigcup t^1(\perp(R_2,\rho)) \bigcup t^1(\perp(R_3,\rho)) \\ r^1(\| R_1) \quad \bigcup \quad r^1(\| R_2) \quad \bigcup \quad r^1(\| R_3) \end{bmatrix}$$
$$= \begin{bmatrix} t^1(\perp(R_1,\rho)) \bigcup t^1(\perp(R_2,\rho)) \bigcup t^1(\perp(R_3,\rho)) \\ r^3 \end{bmatrix}$$

where, r^3—the end link has three finite rotations.

(5) Determine translational elements of the POC set.

Based on the topological characteristics $(SOC\{-\widehat{R_1R_2R_3}-\})$ and Eq. (4.4e, f), the three derivative translations are perpendicular to the same radius vector ρ. It means these three derivative translations are coplanar. So there are only two independent derivative translations. Then the above equation can be further rewritten as

$$M_S = \begin{bmatrix} t^1(\perp(R_1,\rho)) \bigcup t^1(\perp(R_2,\rho)) \bigcup t^1(\perp(R_3,\rho)) \\ r^3 \end{bmatrix} = \begin{bmatrix} t^2(\perp\rho) \\ r^3 \end{bmatrix}$$

where, $t^2(\perp\rho)$—the end link has two finite translations in directions within a plane perpendicular to radius vector ρ.

(6) Select independent elements of the POC set

DOF of the mechanism is three, so there may be up to three independent elements in the POC set.

Therefore, this POC set may be

$$M_S = \begin{bmatrix} \{t^2(\perp\rho)\} \\ r^3 \end{bmatrix} \quad \text{or} \quad M_S = \begin{bmatrix} t^2(\perp\rho) \\ \{r^1(\| R_1)\} \bigcup \{r^1(\| R_2)\} \bigcup r^1(\| R_3) \end{bmatrix}$$

Since DOF of the mechanism is three and the POC set has five elements, only three elements are independent. Two non-independent elements are put into braces. There is $\dim\{M_S\} = 3 = \text{DOF}$.

(7) Discussion

If intersection point o of the three R pairs' axes is selected as the base point, the POC equation will

$$M_S = \begin{bmatrix} t^0 \quad \bigcup \quad t^0 \quad \bigcup \quad t^0 \\ r^1(\| R_1) \bigcup r^1(\| R_2) \bigcup r^1(\| R_3) \end{bmatrix} = \begin{bmatrix} t^0 \\ r^3 \end{bmatrix}$$

where, t^0—the end link has no finite translation, r^3—the end link has three independent finite rotations.

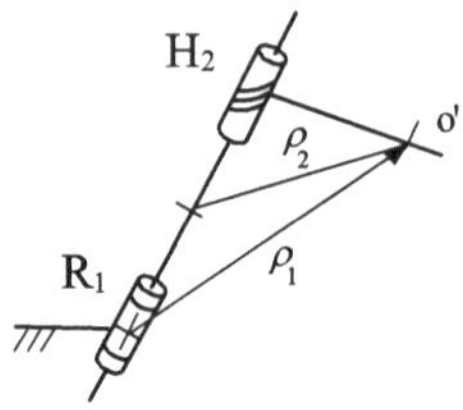

Fig. 4.4 $SOC\{-R_1|H_2-\}$

Example 4.3 Determine POC set of the serial mechanism in Fig. 4.4.

(1) Determine topological structure of the mechanism:

$$SOC\{-R_1|H_2-\}.$$

(2) Select point o′ outside the axis of the end pair (H_2) as base point, as shown in Fig. 4.4.

(3) Establish POC equation.

Substitute POC set of each pair (Eqs. (3.2) and (3.3)) into Eq. (4.4), there is

$$\begin{aligned} M_S &= \begin{bmatrix} t^1(\perp(R_1,\rho_1)) \\ r^1(\| R_1)) \end{bmatrix} \bigcup \begin{bmatrix} t^1(\| H_2) \quad t^1(\perp(H_2,\rho_2)) \\ r^1(\| H_2) \end{bmatrix} \\ &= \begin{bmatrix} t^1(\perp(R_1,\rho_1)) \bigcup t^1(\| H_2) \bigcup t^1(\perp(H_2,\rho_2)) \\ r^1(\| R_1) \quad \bigcup \quad r^1(\| H_2) \end{bmatrix} \end{aligned}$$

(4) Determine rotational elements of the POC set.

Based on the topological structure $(R_1|H_2)$, and the operation rules Eqs. (4.4a) and (4.4j), the above equation can be rewritten as

$$\begin{aligned} M_S &= \begin{bmatrix} t^1(\perp(R_1,\rho_1)) \bigcup t^1(\| H_2) \bigcup t^1(\perp(H_2,\rho_2)) \\ r^1(\| R_1) \quad \bigcup \quad r^1(\| H_2) \end{bmatrix} \\ &= \begin{bmatrix} t^1(\perp(R_1,\rho_1)) \bigcup t^1(\| H_2) \bigcup t^1(\perp(H_2,\rho_2)) \\ r^1(\| R_1) \end{bmatrix} \end{aligned}$$

where, $r^1(\| R_1)$—the end link has a finite rotation in a direction parallel to axis of R_1.

(5) Determine translational elements of the POC set.

Based on the topological characteristics $(R_1|H_2)$, i.e. R_1, H_2, ρ_1 and ρ_2 are coplanar, and the operation rules Eqs. (4.4a) and (4.4j), the above equation can be rewritten as

$$M_S = \begin{bmatrix} t^1(\perp(R_1,\rho_1))\bigcup t^1(\parallel H_2)\bigcup t^1(\perp(H_2,\rho_2)) \\ r^1(\parallel R_1) \end{bmatrix} = \begin{bmatrix} t^1(\perp(R_1)) \quad t^1(\parallel H_2) \\ r^1(\parallel R_1) \end{bmatrix}$$

where, $t^1(\perp R_1)$—the end link has a finite translation in a direction within a plane perpendicular to axis of R_1, $t^1(\parallel H_2)$—the end link has a finite translation in a direction parallel to axis of H_2.

(6) Select independent elements of the POC set

DOF of the mechanism is two, so there may be up to two independent elements in the POC set.

Therefore, this POC set may be

$$M_S = \begin{bmatrix} \{t^1(\perp R_1\}\bigcup t^1(\parallel H_2) \\ r^1(\parallel R_1) \end{bmatrix} \quad \text{or} \quad M_S = \begin{bmatrix} t^1(\perp R_1)\bigcup t^1(\parallel H_2) \\ \{r^1(\parallel R_1)\} \end{bmatrix}$$

Since DOF of the mechanism is two and the POC set has three elements, only two elements are independent. The non-independent element is put into a brace. There is $\dim\{M_S\} = 2 = \text{DOF}$.

(7) Discussion

If a point in axis of H_2 is selected as the base point o′, POC set of the mechanism will be

$$M_S = \begin{bmatrix} t^1(\parallel H_2) \\ r^1(\parallel R_1) \end{bmatrix}$$

where, $r^1(\parallel R_1)$—the end link has a finite rotation in a direction parallel to axis of R_1, $t^1(\parallel H_2)$—the end link has an finite translation in a direction parallel to axis of H_2.

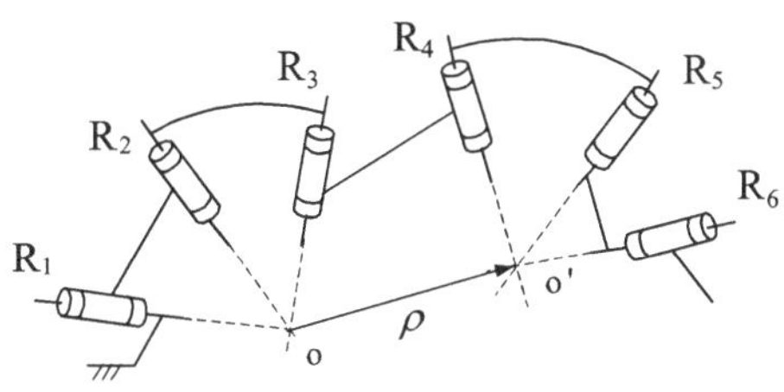

Fig. 4.5 $SOC\{-\widehat{RRR}-\widehat{RRR}-\}$

Example 4.4 Determine POC set of the serial mechanism in Fig. 4.5.

(1) Determine topological structure of the mechanism:

$$SOC\{-\widehat{R_1R_2R_3}-\widehat{R_4R_5R_6}-\}.$$

It is composed of two sub-SOCs (sub-$SOC\{-\widehat{RRR}-\}$ in Table 4.1) connected in series.

(2) Select intersection point (o′) of the last three R pairs as base point, as shown in Fig. 4.5.

(3) Establish POC equation.

Substitute POC set of sub-$SOC\{-\widehat{RRR}-\}$ (No. 6 and No. 6* in Table 4.1) into Eq. (4.4), there is

$$M_S = \begin{bmatrix} t^2(\perp\rho) \\ r^3 \end{bmatrix} \bigcup \begin{bmatrix} t^0 \\ r^3 \end{bmatrix} = \begin{bmatrix} t^2(\perp\rho) \bigcup t^0 \\ r^3 \quad \bigcup r^3 \end{bmatrix}$$

(4) Determine rotational elements of the POC set.

Based on the operation rule Eq. (4.4d), the above equation can be rewritten as

$$M_S = \begin{bmatrix} t^2(\perp\rho) \bigcup t^0 \\ r^3 \quad \bigcup r^3 \end{bmatrix} = \begin{bmatrix} t^2(\perp\rho) \bigcup t^0 \\ r^3 \end{bmatrix}$$

where, r^3—the end link has three finite rotations.

(5) Determine translational elements of the POC set.

The above equation can be further rewritten as

$$M_S = \begin{bmatrix} t^2(\perp\rho) \bigcup t^0 \\ r^3 \end{bmatrix} = \begin{bmatrix} t^2(\perp\rho) \\ r^3 \end{bmatrix}$$

where, $t^2(\perp\rho)$—the end link has two finite translations in directions within a plane perpendicular to radius vector ρ.

(6) Select independent elements of the POC set

DOF of the mechanism is six. According to Eq. (4.4i), there are five independent elements in the POC set.

Therefore, this POC set is

$$M_S = \begin{bmatrix} t^2(\perp\rho) \\ r^3 \end{bmatrix}$$

Since DOF of the mechanism is six and the POC set has five independent elements. There is $\dim\{M_S\} = \dim\{M(r)\} + \dim\{M(t)\} = 3 + 2 = 5 < \mathrm{DOF} = 6$.

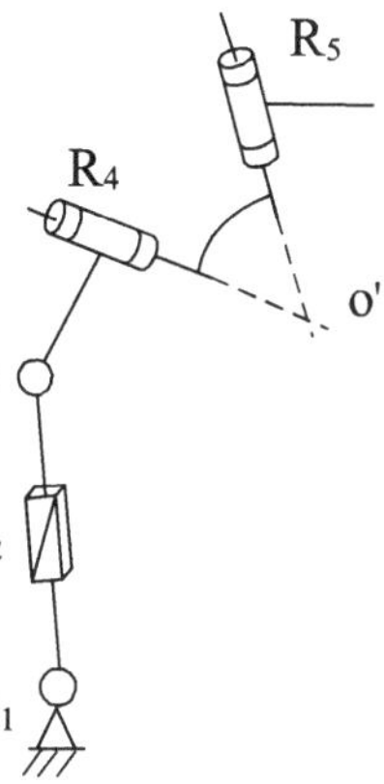

Fig. 4.6 $SOC\{-R(\perp P) \parallel R - \widehat{RR} -\}$

Example 4.5 Determine POC set of the serial mechanism in Fig. 4.6.

(1) Determine topological structure of the mechanism:

$$SOC\{-R_1(\perp P_2) \parallel R_3 - \widehat{R_4R_5} -\}.$$

It is composed of sub-$SOC\{-R(\perp P) \parallel R-\}$ and $SOC\{-\widehat{RR}-\}$ (refer to Table 4.1) connected in series.

(2) Select intersection point (o′) of the axes of R_4 and R_5 as base point, as shown in Fig. 4.6.

(3) Establish POC equation.

Substitute POC set of sub-$SOC\{-R\ (\perp P) \parallel R-\}$ and POC set of sub-$SOC\{-\widehat{RR}-\}$ (No. 2 and No. 5* in Table 4.1) into Eq. (4.4), there is

$$M_S = \begin{bmatrix} t^2(\perp R_1) \\ r^1(\parallel R_1) \end{bmatrix} \bigcup \begin{bmatrix} t^0 \\ r^2(\parallel \diamondsuit(R_4, R_5)) \end{bmatrix} = \begin{bmatrix} t^2(\perp R_1) \bigcup & t^0 \\ r^1(\parallel R_1) \bigcup & r^2(\parallel \diamondsuit(R_4, R_5)) \end{bmatrix}$$

where, $t^2(\perp R_1)$—the end link has two finite translations in directions within a plane perpendicular to axis of R_1.

(4) Determine rotational elements of the POC set.

Based on the topological structure $(R_1 \parallel R_3 \nparallel R_4,)$ and the operation rule Eq. (4.4b), the above equation can be rewritten as

$$M_S = \begin{bmatrix} t^2(\perp R_1) \bigcup & t^0 \\ r^1(\parallel R_1) \bigcup & r^2(\parallel \diamondsuit(R_4, R_5)) \end{bmatrix} = \begin{bmatrix} t^2(\perp R_1) \bigcup t^0 \\ r^3 \end{bmatrix}$$

where, r^3—the end link has three finite rotations.

(5) Determine translational elements of the POC set.

The above equation can be further rewritten as

$$M_S = \begin{bmatrix} t^2(\perp R_1) \bigcup t^0 \\ r^3 \end{bmatrix} = \begin{bmatrix} t^2(\perp R_1) \\ r^3 \end{bmatrix}$$

(6) Select independent elements of the POC set

DOF of the mechanism is five. According to Eq. (4.4i), there are five independent elements in the POC set.

Therefore, this POC set is

$$M_S = \begin{bmatrix} t^2(\perp R_1) \\ r^3 \end{bmatrix}$$

Since DOF of the mechanism is five and the POC set has five independent elements. There is $\dim\{M_S\} = \dim\{M(r)\} + \dim\{M(t)\} = 3 + 2 = 5 =$ DOF.

(7) Discussion

If an arbitrary point o′ outside axis of the end pair R_5 is selected as the base point o′, POC set of the end link will be

$$M_S = \begin{bmatrix} t^2(\perp R_1) & \bigcup \{t^2(\perp \rho)\} \\ r^1(\| R_1) \bigcup & r^2(\| \diamond(R_4, R_5)) \end{bmatrix} = \begin{bmatrix} t^2(\perp R_1) \bigcup \{t^1(\perp \rho)\} \\ r^3 \end{bmatrix}$$

Since DOF of the mechanism is five, the POC set has five independent elements. Another non-independent element is put into a brace.

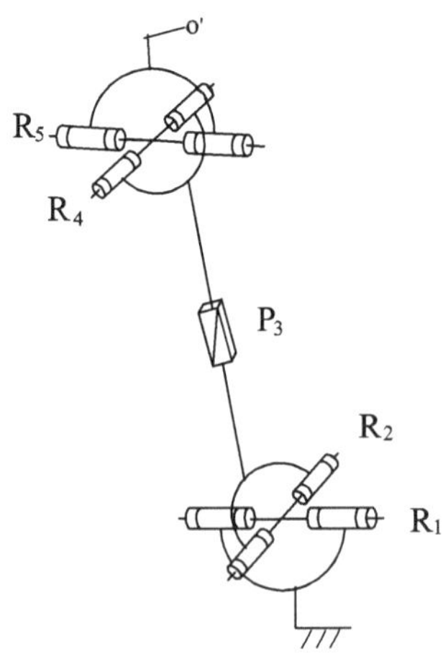

Fig. 4.7 $SOC\{-\widehat{R\perp R}(\perp P) \| \widehat{R\perp R}-\}$

Example 4.6 Determine POC set of the serial mechanism in Fig. 4.7.

(1) Determine topological structure of the mechanism:

$SOC\{-\widehat{R_1 \perp R_2}(\perp P_3) \parallel \widehat{R_4 \perp R_5}-\}$ ($R_2 \perp P_3$ and $R_2 \parallel R_4$; $R_1 \parallel R_5$ upon assembling).

It is composed of two sub-SOCs (sub-$SOC\{-\widehat{R \perp R}-\}$) (refer to Table 4.1) and a P pair connected in series.

(2) Select an arbitrary point o′ outside the axis of R_5 as base point, as shown in Fig. 4.7.

(3) Establish POC equation.

Substitute POC set of sub-$SOC\{-\widehat{R \perp R}-\}$ (No. 5 in Table 4.1) and POC set of P pair (Eq. (3.1)) into Eq. (4.4), there is

$$M_S = \begin{bmatrix} t^2(\perp \rho_1) \\ r^2(\parallel \diamondsuit(R_1, R_2)) \end{bmatrix} \bigcup \begin{bmatrix} t^1(\parallel P_3) \\ r^0 \end{bmatrix} \bigcup \begin{bmatrix} t^2(\perp \rho_2) \\ r^2(\parallel \diamondsuit(R_4, R_5)) \end{bmatrix}$$
$$= \begin{bmatrix} t^2(\perp \rho_1) & \bigcup t^1(\parallel P_3) \bigcup & t^2(\perp \rho_2) \\ r^2(\parallel \diamondsuit(R_1, R_2)) & \bigcup \quad r^0 \quad \bigcup & r^2(\parallel \diamondsuit(R_4, R_5)) \end{bmatrix}$$

(4) Determine rotational elements of the POC set.

Based on the topological structure ($R_1 \parallel R_5$ and $R_2 \parallel R_4$,) and Eq. (4.4c), the above equation can be rewritten as

$$M_S = \begin{bmatrix} t^2(\perp \rho_1) & \bigcup t^1(\parallel P_3) \bigcup & t^2(\perp \rho_2) \\ r^2(\parallel \diamondsuit(R_1, R_2)) & \bigcup \quad r^0 \quad \bigcup & r^2(\parallel \diamondsuit(R_4, R_5)) \end{bmatrix}$$
$$= \begin{bmatrix} t^2(\perp \rho_1) \bigcup t^1(\parallel P_3) \bigcup t^2(\perp \rho_2) \\ r^2(\parallel \diamondsuit(R_1, R_2)) \end{bmatrix}$$

where, $r^2(\parallel \diamondsuit(R_1, R_2))$—the end link has two finite rotations in directions within a plane parallel to axes of R_1 and R_2 .

(5) Determine translational elements of the POC set.

According to operation rule Eq. (4.4f, h), the above equation can be further rewritten as

$$M_S = \begin{bmatrix} t^2(\perp \rho_1) \bigcup t^1(\parallel P_3) \bigcup t^2(\perp \rho_2) \\ r^2(\parallel \diamondsuit(R_1, R_2)) \end{bmatrix} = \begin{bmatrix} t^3 \\ r^2(\parallel \diamondsuit(R_1, R_2)) \end{bmatrix}$$

where, t^3—the end link has three finite translations.

(6) Select independent elements of the POC set

DOF of the mechanism is five. According to Eq. (4.4i), there are five independent elements in the POC set.

Therefore, this POC set is

$$M_S = \begin{bmatrix} t^3 \\ r^2(\| \diamondsuit(R_1, R_2)) \end{bmatrix}$$

Since DOF of the mechanism is five and the POC set has five independent elements. There is $\dim\{M_S\} = \dim\{M(r)\} + \dim\{M(t)\} = 2 + 3 = 5 = \text{DOF}$.

It should be noted that $R_1 \parallel R_5$ is only an instantaneous topological structure when assembling. POC set of end link is also an instantaneous one.

4.6 Number of Independent Displacement Equations of Single-Loop Mechanisms

(1) Single-loop chain and corresponding SOC

If a certain link of a single-loop chain (abbreviated as SLC) is broken, an SOC corresponding to this SLC can be obtained (written as $\text{SOC}_{(\text{SLC})}$). The link to break should be so selected that the SLC and the $\text{SOC}_{(\text{SLC})}$ must have the same topological structure [17, 18].

For example, for the SLC in Fig. 4.8a, if the link between R_1 and R_6 is broken, the obtained $\text{SOC}_{(\text{SLC})}$ and the SLC have the same topological structure (including sub-$SOC\{-R_1 \parallel R_2 \parallel R_3-\}$ and sub-$SOC\{-\widehat{R_4R_5R_6}-\}$ connected in series). If the link between R_5 and R_6 is broken, the obtained $\text{SOC}_{(\text{SLC})}$ and the SLC have different topological structures. So, the link between R_5 and R_6 should not be selected as the link to break.

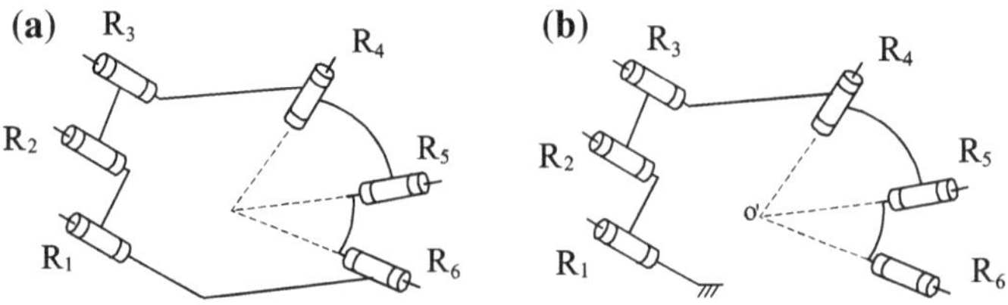

Fig. 4.8 $SLC\{-R_1 \parallel R_2 \parallel R_3 - \widehat{R_4R_5R_6}-\}$

(2) Number of independent displacement equations of an SLC

In order to obtain the $SOC_{(SLC)}$ corresponding to the SLC, there may be several links which meet the above topological structure equivalence condition. Among these links, the one should be so selected that POC set of the resultant $SOC_{(SLC)}$ has the number of non-independent elements as few as possible. Then number of independent displacement equations of the SLC is equal to dimension of POC set of the corresponding $SOC_{(SLC)}$ [17, 18], i.e.

$$\xi_L = \dim\{M_{S(L)}\} \tag{4.5}$$

where, ξ_L—number of independent displacement equations of the SLC, $\dim\{M_{S(L)}\}$—dimension of POC set of the corresponding $SOC_{(SLC)}$.

Physical meaning of Eq. (4.5): In order to restrict relative motions between two end links of an $SOC_{(SLC)}$, ξ_L ($\dim\{M_{S(L)}\}$) constraints can be attached between the two end links to make them into an integral one. And thus an SLC is obtained.

Example 4.7 Determine number of independent displacement equations (ξ_L) of the single-loop mechanism in Fig. 4.8a.

(1) Determine topological structure of the mechanism:

$$SLC\{-R_1 \parallel R_2 \parallel R_3 - \overbrace{R_4R_5R_6}-\}.$$

It is composed of sub-$SOC\{-R \parallel R \parallel R-\}$ and sub-$SOC\{-\overbrace{RRR}-\}$ (refer to Table 4.1) connected in series.

(2) Suppose the link between R_1 and R_6 is broken (Fig. 4.8a) and select the intersect point (o′) of axes of R_4, R_5 and R_6 as base point, as shown in Fig. 4.8b.

(3) Establish POC equation.

Substitute POC set of sub-$SOC\{-R \parallel R \parallel R-\}$ and POC set of sub-$SOC\{-\overbrace{RRR}-\}$(No. 2 and No. 6* in Table 4.1) into Eq. (4.4), there is

$$M_{S(L)} = \begin{bmatrix} t^2(\perp R_3) \\ r^1(\parallel R_3) \end{bmatrix} \bigcup \begin{bmatrix} t^0 \\ r^3 \end{bmatrix} = \begin{bmatrix} t^2(\perp R_3) \bigcup t^0 \\ r^1(\parallel R_3) \bigcup r^3 \end{bmatrix}$$

(4) Determine rotational elements of the POC set.

According to operation rule Eq. (4.4d), the above equation can be rewritten as

$$M_{S(L)} = \begin{bmatrix} t^2(\perp R_3) \\ r^1(\parallel R_3) \end{bmatrix} \bigcup \begin{bmatrix} t^0 \\ r^3 \end{bmatrix} = \begin{bmatrix} t^2(\perp R_3) \bigcup t^0 \\ r^3 \end{bmatrix}$$

(5) Determine translational elements of the POC set.

The above equation can be further rewritten as

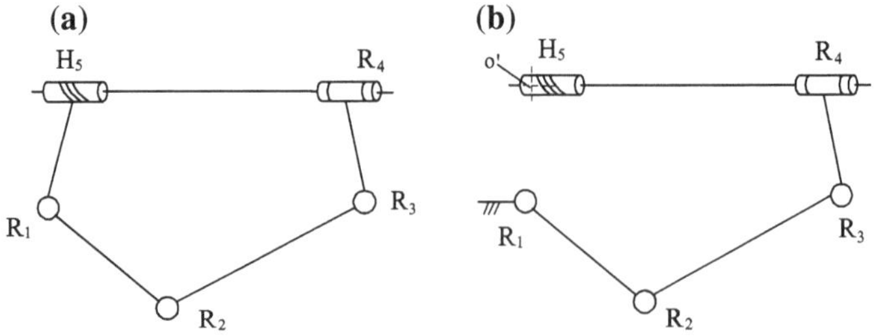

Fig. 4.9 $SLC\{-R_1 \parallel R_2 \parallel R_3 - R_4|H_5-\}$

$$M_{S(L)} = \begin{bmatrix} t^2(\perp R_3) \bigcup t^0 \\ r^3 \end{bmatrix} = \begin{bmatrix} t^2(\perp R_3) \\ r^3 \end{bmatrix}$$

(6) Determine number of independent displacement equations

Since DOF of the mechanism is five and the POC set has five independent elements. There is $\dim\{M_{S(L)}\} = \dim\{M(r)\} + \dim\{M(t)\} = 3 + 2 = 5$. According to Eq. (4.5), number of independent displacement equations (ξ_L) of the SLC in Fig. 4.8a is $\xi_L = \dim\{M_{S(L)}\} = 5$.

Example 4.8 Determine number of independent displacement equations (ξ_L) of the single-loop mechanism in Fig. 4.9a.

(1) Determine topological structure of the mechanism:

$$SLC\{-R_1 \parallel R_2 \parallel R_3 - R_4|H_5-\}.$$

It is composed of sub-$SOC\{-R \parallel R \parallel R-\}$ and sub-$SOC\{-R|H-\}$ (refer to Table 4.1) connected in series.

(2) Suppose the link between R_1 and H_5 is broken (Fig. 4.9a) and select a point (o′) on axis of H_5 as base point, as shown in Fig. 4.9b.

(3) Establish POC equation.

Substitute POC set of sub-$SOC\{-R \parallel R \parallel R-\}$ and POC set of sub-$SOC\{-R|H-\}$ (No. 2 and No. 7* in Table 4.1) into Eq. (4.4), there is

$$M_{S(L)} = \begin{bmatrix} t^2(\perp R_3) \\ r^1(\parallel R_3) \end{bmatrix} \bigcup \begin{bmatrix} t^1(\parallel H_5) \\ r^1(\parallel R_4) \end{bmatrix} = \begin{bmatrix} t^2(\perp R_3) \bigcup t^1(\parallel H_5) \\ r^1(\parallel R_3) \bigcup r^1(\parallel R_4) \end{bmatrix}$$

(4) Determine rotational elements of the POC set.

Based on topological structure ($R_3 \nparallel R_4$) and operation rule Eq. (4.4a), the above equation can be rewritten as

$$M_{S(L)} = \begin{bmatrix} t^2(\perp R_3) \bigcup t^1(\| H_5) \\ r^1(\| R_3) \bigcup r^1(\| R_4) \end{bmatrix} = \begin{bmatrix} t^2(\perp R_3) \bigcup t^1(\| H_5) \\ r^2(\| \diamondsuit(R_3, R_4)) \end{bmatrix}$$

(5) Determine translational elements of the POC set.

According to Eq. (4.4f), the above equation can be further rewritten as

$$M_{S(L)} = \begin{bmatrix} t^2(\perp R_3) \bigcup t^1(\| H_5) \\ r^2(\| \diamondsuit(R_3, R_4) \end{bmatrix}$$
$$= \begin{cases} \begin{bmatrix} t^2(\perp R_3) \\ r^2(\| \diamondsuit(R_3, R_4)) \end{bmatrix}, & if\ R_3 \perp R_4; \\ \begin{bmatrix} t^3 \\ r^2(\| \diamondsuit(R_3, R_4)) \end{bmatrix}, & if\ R_3 \not\perp R_4. \end{cases}$$

(6) Determine number of independent displacement equations

Since DOF of the mechanism is five, number of independent displacement equations (ξ_L) of the SLC in Fig. 4.9a is

$$\xi_L = \dim\{M_{S(L)}\} = \begin{cases} 4, & if\ R_3 \perp R_4; \\ 5, & if\ R_3 \not\perp R_4. \end{cases}$$

4.7 Summary

(1) Based on topological structure invariance of serial mechanisms (excluding singular position) and unit vector set of the end link velocity, VC equation for serial mechanisms (Eq. 4.3) and the corresponding symbolic operation rules are derived.

(2) Based on invariance property of the end link VC set and one-to-one correspondence between elements of VC set and elements of POC set, POC equation for serial mechanisms (Eq. 4.4) and the corresponding symbolic operation rules are obtained.

(3) Different types of common sub-SOCs contained in mechanisms and their POC sets are listed in Table 4.1. They may help to simplify generation and calculation of POC equation for serial mechanisms.

(4) The POC equation for serial mechanisms (Eq. 4.4) can be used to determine number of independent displacement equations of single-loop mechanisms (Eq. 4.5).

(5) The POC equation for serial mechanisms (Eq. 4.4) and corresponding operation rules are independent of the mechanism motion position and the fixed coordinate system (i.e. it is not necessary to establish the fixed coordinate system). So, it could be called as a geometrical method.

(6) The POC equation for serial mechanisms can be used in structure synthesis of serial mechanisms (Chap. 8) and establishment of POC equation for parallel mechanisms (Chap. 5).

References

1. Hunt KH (1987) Kinematic geometry of mechanisms. Clarendon Press, Oxford
2. Huang Z, Li QC (2002) General methodology for type synthesis of symmetrical lower-mobility parallel manipulators and several novel manipulators. Int J Robot Res 21:131–145
3. Kong X, Gosselin C (2007) Type synthesis of parallel mechanisms. Springer, Heidelberg
4. Meng X, Gao F, Shengfu WuS, Ge QJ (2014) Type synthesis of parallel robotic mechanisms: framework and brief review. Mech Mach Theor 78:177–186
5. Herve JM (1978) Analyse structurelle des mécanismes par groupe des déplacements. Mech Mach Theor 13:437–450
6. Fanghella P, Galletti C (1995) Metric relations and displacement groups in mechanism and robot kinematics. ASME J Mech Des 117:470–478
7. Herve JM (1999) The lie group of rigid body displacements, a fundamental tool for mechanism design. Mech Mach Theor 34:719–730
8. Meng J, Liu GF, Li ZX (2007) A geometric theory for analysis and synthesis of sub-6 DOF parallel manipulators. IEEE Trans Robot 23:625–649
9. Gogu G (2007) Structural synthesis of parallel robots: part 1: methodology. Springer, Dordrecht
10. Yang T-L, Jin Q et al (2001) A general method for type synthesis of rank-deficient parallel robot mechanisms based on SOC Unit. Mach Sci Technol 20(3):321–325
11. Yang T-L, Jin Q, Liu A-X, Shen H-P, Luo Y-F (2002) Structural synthesis and classification of the 3-DOF translation parallel robot mechanisms based on the unites of single-open chain. Chin J Mech Eng 38(8):31–36. doi:10.3901/JME.2002.08.031
12. Jin Q, Yang T-L (2004) Theory for topology synthesis of parallel manipulators and its application to three-dimension-translation parallel manipulators. ASME J Mech Des 126: 625–639
13. Yang T-L (2004) Theory of topological structure for robot mechanisms. China Machine Press, Beijing
14. Yang T-L, Liu A-X, Luo Y-F, Shen H-P et al (2009) Position and orientation characteristic equation for topological design of robot mechanisms. ASME J Mech Des 131:021001-1–021001-17
15. Yang T-L, Liu A-X, Luo Y-F, Shen H-P et al (2013) On the correctness and strictness of the position and orientation characteristic equation for topological design of robot mechanisms. ASME J Mech Robot 5:021009-1–021009-18
16. Yang T-L, Liu A-X, Shen H-P et al (2012) Theory and application of robot mechanism topology. China Science Press, Beijing
17. Yang T-L, Sun D-J (2008) Rank and mobility of single loop kinematic chains. In: Proceedings of the ASME 32-th mechanisms and robots conference, DETC2008-49076
18. Yang T-L, Sun D-J (2012) A general DOF formula for parallel mechanisms and multi-loop spatial mechanisms. ASME J Mech Robot 4(1):011001-1–011001-17
19. Yang T-L, Liu A-X, Shen H-P et al (2015) Composition principle based on SOC unit and coupling degree of BKC for general spatial mechanisms. In: The 14th IFToMM world congress, Taipei, Taiwan, 25–30 Oct 2015. doi: 10.6567/IFToMM.14TH.WC.OS13.135

Chapter 5
Position and Orientation Characteristics Equation for Parallel Mechanisms

5.1 Introduction

POC equation for parallel mechanisms and the corresponding operation rules are established in this chapter. The POC equation for parallel mechanisms reveals the mapping relations among topological structure, POC set and DOF of parallel mechanisms.

For a parallel mechanism whose topological structure is known, there are the following four types of methods to determine the relative motion characteristics of the moving platform to the fixed platform.

(1) Method based on screw theory

   Motion screw system of a branch is converted into constraint screw system through reciprocal product of motion systems. Constraint screw system of the moving platform is obtained through "union" operation of each branch's constraint screw system and then is converted into motion screw system of the moving platform again through reciprocal product of screw systems. This method depends on motion position of the mechanism and the fixed coordinate system. But it avoids "intersection" operation of motion screw systems [1, 2].

(2) Method based on displacement subgroup/submanifold

   For PMs fulfilling algebra structures of Lie group, relative motion characteristics of the moving platform can be obtained through "intersection" operation rules of displacement subgroups using synthetic arguments or tables of compositions of subgroups, which is independent of motion position of a mechanism [2–5]. For PMs not fulfilling the algebra structures of Lie group, its motion characteristics can be obtained by the method based on displacement submanifold. But this method depends on motion position of mechanisms [5].

T.-L. Yang et al., *Topology Design of Robot Mechanisms*,
https://doi.org/10.1007/978-981-10-5532-4_5

(3) Method based on linear transformation

The moving platform's velocity space relative to the fixed platform can be obtained through "intersection" operation of the velocity space of each branch's end link relative to the branch's fixed link. It can be determined by intuition [2, 6]. This method depends on motion position of the mechanism and the fixed coordinate system.

(4) Method based on POC equation for parallel mechanisms

POC set of the moving platform can be determined by POC equation for parallel mechanisms (Eq. 5.3) [7–15]. This method is a geometrical method. It is independent of motion position (excluding singular positions) of the mechanism and the fixed coordinate system (i.e. it is not necessary to establish the fixed coordinate system).

This chapter focuses on establishment of POC equation for parallel mechanisms and corresponding operation rules [7–11]. Firstly, VC equation for parallel mechanisms and corresponding operation rules are established based on (a) invariance of parallel mechanism topological structure and (b) unit vector set of the moving platform velocity being "intersection" of unit vector set of each branch end link's velocity. Then based on invariance property of VC set and one-to-one correspondence between elements of VC set and elements of POC set, POC equation for parallel mechanisms and corresponding operation rules are obtained.

5.2 VC Equation for Parallel Mechanisms and Its Operation Rules

5.2.1 An Introductory Example

In order to describe the basic idea for establishment of VC equation for parallel mechanisms, an introductory example is given as below.

Introductory example Determine VC set and POC set of the moving platform of the parallel mechanism in Fig. 5.1.

(1) Determine topological structure of the mechanism:

- Topological structure of branches:

$$SOC\{-R_{j1} \parallel R_{j2} \parallel C_{j3}-\}, \quad j = 1, 2, 3$$

- Topological structure of two platforms:

Axes of R_{11}, R_{21} and R_{31} form a triangle, as shown in Fig. 5.1.

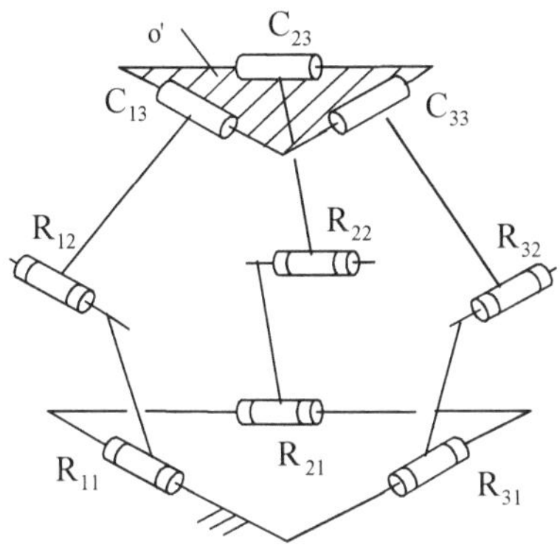

Fig. 5.1 A 3-translation parallel mechanism

(2) Select an arbitrary point (o') on the moving platform as base point, as shown in Fig. 5.1.

(3) Determine VC set of the end link of each branch.

For the same base point o′, VC set of the end link of each branch can be obtained based on VC equation for serial mechanisms (Eq. (4.3) in Chap. 4).

$$\dot{M}_{bj} = \begin{bmatrix} \dot{t}^3 \\ \dot{r}^1(\| R_{j1}) \end{bmatrix}, \quad \text{j} = 1, 2, 3.$$

According to Chap. 4, VC set of the end link of each branch depends only on topological structure of the branch and is independent of motion position of the mechanism (excluding singular position) and the fixed coordinate system (refer to Sect. 4.3.1 for detail).

(4) Determine VC set of the moving platform of a parallel mechanism.

The unit vector set of velocity of moving platform of this PM is "intersection" of the unit vector sets of velocity of each branch's end link, which depends only on its topological structure. Hence, this unit vector set has the invariance. Since this unit vector set could be rewritten as the VC set (refer to Chap. 3), the VC set of moving platform of this PM is "intersection" of the VC sets of each branch's end link. And the VC set also has invariance.

Therefore, the VC set of moving platform of the PM in Fig. 5.1 can be written as

$$\dot{M}_{Pa} = \bigcap_{j=1}^{3} \dot{M}_{bj} = \begin{bmatrix} \dot{t}^3 \\ \dot{r}^1(\| R_{11}) \end{bmatrix} \bigcap \begin{bmatrix} \dot{t}^3 \\ \dot{r}^1(\| R_{21}) \end{bmatrix} \bigcap \begin{bmatrix} \dot{t}^3 \\ \dot{r}^1(\| R_{31}) \end{bmatrix}$$

(a) Determine rotational elements of this VC set

The end link of each branch has a rotational element, but these three rotation elements are not parallel each other. So, the moving platform has no rotation. There is

$$\left[\dot{r}^1(\| R_{11})\right]\bigcap\left[\dot{r}^1(\| R_{21})\right]\bigcap\left[\dot{r}^1(\| R_{31})\right] = \left[\dot{r}^0\right]$$

(b) Determine translational elements of this VC set
Since the end link of each branch has three translational elements, "intersection" of the VC set of each branch's end link also has three translational elements. There is

$$\left[\dot{t}^3\right]\bigcap\left[\dot{t}^3\right]\bigcap\left[\dot{t}^3\right] = \left[\dot{t}^3\right]$$

Hence, the VC set of the moving platform of the parallel mechanism in Fig. 5.1 is

$$\dot{M}_{Pa} = \bigcap_{j=1}^{3}\dot{M}_{bj} = \begin{bmatrix} \dot{t}^3 \\ \dot{r}^0 \end{bmatrix}$$

Since the mechanism topological structure has invariance property (Chap. 2), VC set of the moving platform also has invariance property.

(5) Determine POC set of the moving platform.
Since VC set of the moving platform has invariance property (excluding singular position). It means that each element of VC set corresponds to a finite displacement. Based on the above VC set, moving platform of the parallel mechanism in Fig. 5.1 has no finite rotation and three finite translations. Hence, POC set of the moving platform can be written as

$$M_{Pa} = \begin{bmatrix} t^3 \\ r^0 \end{bmatrix}$$

5.2.2 *VC Equation for Parallel Mechanisms*

A parallel mechanism with v-loop can be regarded as being formed by moving platform, fixed platform and $(v+1)$ SOCs parallels connected between the two platforms, as shown in Fig. 5.1. Motion of the moving platform is restrained by these $(v+1)$ SOCs.

According to the mechanics principle and the introductory example in Sect. 5.2.1:

(1) The velocity of moving platform of this PM is "intersection" of velocities of each branch's end link, which depends on the motion position and the fixed coordinate system.

(2) The unit vector set of moving platform velocity of this PM is "intersection" of the unit vector sets of each branch's end link velocity and could be written as

$$V_{Pa}^{e} = \bigcap_{j=1}^{(v+1)} V_{b_j}^{e} \tag{5.1}$$

where, V_{Pa}^{e}—unit vector set of the moving platform velocity (refer to Eq. (3.7)); $V_{b_j}^{e}$—unit vector set of velocities of each branch's end link for the same base point o′.

Equation (5.1) depends only on its topological structure. Due to the topological structure invariance (excluding singular positions) of a mechanism described in Chap. 2, the unit vector set of moving platform velocity of the PM has the invariance and is independent of motion position of the PM.

(3) Since the unit vector set of velocity could be rewritten as the VC set described in Chap. 3, the VC set of moving platform of the PM is "intersection" of the VC sets of each branch's end link, which also has the invariance and is independent of motion position of a mechanism (excluding singular positions) and the fixed coordinate system.

Therefore, the VC equation for PMs can be written as [7–11]

$$\dot{M}_{Pa} = \bigcap_{j=1}^{(v+1)} \dot{M}_{b_j} \tag{5.2}$$

where, $\dot{M}_{Pa}$—the VC set of the moving platform relative to the fixed platform of a PM (in short, the VC set of the PM), $\dot{M}_{b_j}$—VC set of the jth branch's end link for the same base point o′.

Equation (5.2) is called VC equation for parallel mechanisms.

It should be noted that:

(a) Number of independent elements in the moving platform's VC set is not greater than DOF of the mechanism (Eq. (3.10) in Chap. 3).

(b) Equation (5.2) does not include VC set of inactive kinematic pair. Inactive pairs in the mechanism should be determined and eliminated (refer to Sect. 6.3.1 and Example 6.2 in Chap. 6 for determination method).

Therefore, calculation of VC set of a parallel mechanism always involves DOF calculation and inactive pair determination.

5.2.3 Operation Rules for VE Equation

For VC equation for parallel mechanisms (Eq. 5.2), there are the following three types of symbolic operation rules [7–11]:

(1) Operation rules for "intersection" of rotational elements [linear operation: Eqs. (5.2a)–(5.2f)].

(2) Operation rules for "intersection" of translational elements [linear operation: Eqs. (5.2g)–(5.2l)].

(3) Rules for selection of independent elements [nonlinear criteria: Eqs. (5.2m), (5.2n)].

A. **Operation rules for "intersection" of rotational elements**

(1) $(\dot{r}^1 \bigcap \dot{r}^3)$:

$$\left[\dot{r}^1(\| R_i)\right]_{b_i} \bigcap \left[\dot{r}^3\right]_{b_j} = \left[\dot{r}^1(\| R_i)\right]_{Pa} \tag{5.2a}$$

where, $\dot{r}^1(\| R_i)$—one rotational element (unit vector of angular velocity) in a direction parallel to axis of R_i.

(2) $(\dot{r}^2 \bigcap \dot{r}^3)$:

$$\left[\dot{r}^2(\| \diamondsuit(R_{i_1}, R_{i_2}))\right]_{b_i} \bigcap \left[\dot{r}^3\right]_{b_j} = \left[\dot{r}^2(\| \diamondsuit(R_{i_1}, R_{i_2}))\right]_{Pa} \tag{5.2b}$$

where, $\dot{r}^2(\| \diamondsuit(R_{i_1}, R_{i_2}))$—two rotational elements (unit vector of angular velocity) in any direction within a plane parallel to axis of R_{i_1} and axis of R_{i_2}, $\dot{r}^3$—three rotational elements (unit vector of angular velocity) in any direction within 3D space.

(3) $(\dot{r}^3 \bigcap \dot{r}^3)$:

$$\left[\dot{r}^3\right]_{b_i} \bigcap \left[\dot{r}^3\right]_{b_j} = \left[\dot{r}^3\right]_{Pa} \tag{5.2c}$$

If each branch contains an $SOC\{-\widehat{RRR}-\}$, Eq. (5.2c) can be rewritten as

$$\left[\dot{r}^3(o_1)\right]_{b_i} \bigcap \left[\dot{r}^3(o_2)\right]_{b_j} = \left[\dot{r}^1(\| (o_1 - o_2)) \bigcup \dot{r}^2\right]_{Pa}$$

where, $o_1(o_2)$—intersection of axes of three R pairs, $(o_1 - o_2)$—the line connecting o_1 and o_2.

(4) $(\dot{r}^1 \bigcap \dot{r}^1)$:

$$\left[\dot{r}^1(\| R_i)\right]_{b_i} \bigcap \left[\dot{r}^1(\| R_j)\right]_{b_j} = \begin{cases} \left[\dot{r}^1(\| R_i)\right]_{Pa}, & \text{if } R_i \| R_j. \\ \left[\dot{r}^0\right]_{Pa}, & \text{if } R_i \nparallel R_j \end{cases} \tag{5.2d}$$

where, $\dot{r}^0$—there is no angular velocity.

(5) $(\dot{r}^1 \bigcap \dot{r}^2)$:

$$\left[\dot{r}^1(\| R_i)\right]_{b_i} \cap \left[\dot{r}^2(\| \Diamond(R_{j_1}, R_{j_2}))\right] = \begin{cases} [\dot{r}^1(\| R_i)]_{Pa}, & \text{if } R_i \| (\Diamond(R_{j_1}, R_{j_2})). \\ [\dot{r}^0]_{Pa}, & \text{if } R_i \nparallel (\Diamond(R_{j_1}, R_{j_2})). \end{cases} \quad (5.2e)$$

(6) $(\dot{r}^2 \bigcap \dot{r}^2)$:

$$\begin{aligned} &[\dot{r}^2(\| \Diamond(R_{i_1}, R_{i_2}))]_{b_i} \bigcap [\dot{r}^2(\| \Diamond(R_{j_1}, R_{j_2}))]_{b_j} \\ &= \begin{cases} [\dot{r}^2(\| \Diamond(R_{i_1}, R_{i_2}))]_{Pa}, & \text{if } (\Diamond(R_{i_1}, R_{i_2})) \| (\Diamond(R_{j_1}, R_{j_2})). \\ \left[\dot{r}^1(\| (\Diamond(R_{i_1}, R_{i_2}) \bigcap (\Diamond(R_{j_1}, R_{j_2})))\right]_{Pa}, & \text{if } (\Diamond(R_{i_1}, R_{i_2})) \nparallel (\Diamond(R_{j_1}, R_{j_2})). \end{cases} \end{aligned} \quad (5.2f)$$

where, $\dot{r}^1(\| (\Diamond(R_{i_1}, R_{i_2}) \bigcap \Diamond(R_{j_1}, R_{j_2})))$—one rotational element (unit vector of angular velocity) in a direction parallel to intersection line of plane $\Diamond(R_{i_1}, R_{i_2})$ and plane $\Diamond(R_{j_1}, R_{j_2})$.

B. **Operation rules for "intersection" of translational elements**

(1) $(\dot{t}^1 \bigcap \dot{t}^3)$:

$$[\dot{t}^1(\| P_i^*)]_{b_i} \bigcap [\dot{t}^3]_{b_j} = [\dot{r}^1(\| P_i^*)]_{Pa} \quad (5.2g)$$

where, P^*—translation of a P pair or derivative translation of an R (H) pair, $\dot{t}^1(\| P_i^*)$—one translational element (unit vector of translational velocity) in a direction parallel to P^*, $\dot{t}^3$—three translational elements (unit vector of translational velocity) in any direction within 3D space.

(2) $(\dot{t}^2 \bigcap \dot{t}^3)$:

$$\left[\dot{t}^2(\| \Diamond(P_{i_1}^*, P_{i_2}^*))\right]_{b_i} \bigcap [\dot{t}^3]_{b_j} = \left[\dot{t}^2(\| \Diamond(P_{i_1}^*, P_{i_2}^*))\right]_{Pa} \quad (5.2h)$$

where, $\dot{t}^2(\| \Diamond(P_{i_1}^*, P_{i_2}^*))$—two translation elements (unit vector of translational velocity) in any direction within a plane parallel to $P_{i_1}^*$ and $P_{i_2}^*$.

(3) $(\dot{t}^3 \bigcap \dot{t}^3)$:

$$[\dot{t}^3]_{b_i} \bigcap [\dot{t}^3]_{b_j} = [\dot{t}^3]_{Pa} \tag{5.2i}$$

(4) $(\dot{t}^1 \bigcap \dot{t}^1)$:

$$[\dot{t}^1(\| P_i^*)]_{b_i} \bigcap \left[\dot{t}^1(\| P_j^*)\right]_{b_j} = \begin{cases} [\dot{t}^1(\| P_i^*)]_{Pa}, & if\ P_i^* | P_j^*. \\ [\dot{t}^0]_{Pa}, & if\ P_i^* \nparallel P_j^*. \end{cases} \tag{5.2j}$$

where, $\dot{t}^0$—there is no translational velocity.

(5) $(\dot{t}^1 \bigcap \dot{t}^2)$:

$$[\dot{t}^1(\| P_i^*)]_{b_i} \bigcap \left[\dot{t}^2(\| \diamondsuit(P_{j_1}^*, P_{j_2}^*))\right] = \begin{cases} [\dot{t}^1(\| P_i^*)]_{Pa}, & if\ P_i^* \parallel (\diamondsuit(P_{j_1}^*, P_{j_2}^*)). \\ [\dot{t}^0]_{Pa}, & if\ P_i^* \nparallel (\diamondsuit(P_{j_1}^*, P_{j_2}^*)). \end{cases} \tag{5.2k}$$

(6) $(\dot{t}^2 \cap \dot{t}^2)$:

$$\begin{aligned} &\left[\dot{t}^2(\| \diamondsuit(P_{i_1}^*, P_{i_2}^*))\right]_{b_i} \bigcap \left[\dot{t}^2(\| \diamondsuit(P_{j_1}^*, P_{j_2}^*))\right] \\ &= \begin{cases} \left[\dot{t}^2(\| \diamondsuit(P_{i_1}^*, P_{i_2}^*))\right]_{Pa}, & if\ (\diamondsuit(P_{i_1}^*, P_{i_2}^*)) \parallel (\diamondsuit(P_{j_1}^*, P_{j_2}^*)). \\ \left[\dot{t}^1(\| (\diamondsuit(P_{i_1}^*, P_{i_2}^*) \bigcap \diamondsuit(P_{j_1}^*, P_{j_2}^*)))\right]_{Pa}, & if\ (\diamondsuit(P_{i_1}^*, P_{i_2}^*)) \nparallel (\diamondsuit(P_{j_1}^*, P_{j_2}^*)). \end{cases} \end{aligned} \tag{5.2l}$$

where, $\dot{t}^1(\| (\diamondsuit(P_{i_1}^*, P_{i_2}^*) \bigcap \diamondsuit(P_{j_1}^*, P_{j_2}^*)))$—one translational element (unit vector of translational velocity) in a direction parallel to intersection line of plane $\diamondsuit(P_{i_1}^*, P_{i_2}^*)$ and plane $\diamondsuit(P_{j_1}^*, P_{j_2}^*)$.

C. **Criteria for selection of independent elements**

(1) Number of independent elements of the moving platform's VC set shall not be greater than DOF of the mechanism, i.e.

$$rank\{\dot{M}_{Pa}\} \leq DOF \tag{5.2m}$$

where, $rank\{\dot{M}_{Pa}\}$—number of independent elements of the moving platform's VC set.

If DOF of the mechanism is zero, the VC set will be

$$\dot{M}_{Pa} = \bigcap_{j=1}^{v+1} \dot{M}_{b_j} = \begin{bmatrix} \dot{t}^0 \\ \dot{r}^0 \end{bmatrix}$$

(2) Any two independent elements can not be from VC set of the same R (or H) pair. (5.2n)

It should be noted that these symbolic operation rules depend on only topological structure and are independent of motion position of the mechanism and it is not necessary to establish the fixed coordinate system.

5.3 POC Equation for Parallel Mechanisms and Its Operation Rules

5.3.1 POC Equation for Parallel Mechanisms

Due to the VC set has invariance (excluding singular positions) described in Sect. 5.2.2, each element of the VC set of Eq. (5.2) corresponds to a finite displacement. Correspondingly, the POC equation for PMs can be written as [7–11]:

$$M_{Pa} = \bigcap_{j=1}^{(v+1)} M_{b_j} \tag{5.3}$$

where, M_{Pa}—POC set of the moving platform relative to the fixed platform of a PM (in short, the POC set of a PM), M_{b_j}—POC set of the jth branch's end link for the same base point o′.

It should be noted that:

(a) Number of independent elements in the moving platform's POC set is not greater than DOF of the mechanism (Eq. 3.10 in Chap. 3).

(b) Equation (5.3) does not include POC set of inactive kinematic pair. Inactive pairs in the mechanism should be determined and eliminated before calculation of POC set (refer to Sect. 6.3.1 and Example 6.2 in Chap. 6 for determination method).

Therefore, calculation of POC set of a parallel mechanism always involves DOF calculation and inactive pair determination.

Equation (5.3) is called the POC equation for PMs, which also has the invariance and is independent of motion position of a mechanism (excluding singular positions) and the fixed coordinate system. It reveals the functional relations among mechanism topological structure, POC set and DOF. Forward operation of this POC equation can be used in structure analysis of PMs, i.e. determining POC set and DOF of a PM when its topological structure is known (refer to Sect. 5.5 and Chap. 6).

Inverse operation of this POC equation can be used in structure synthesis of PMs, i.e. determining topological structure of the PM when its POC set and DOF are known (Chap. 9).

5.3.2 Operation Rules for POC Equation

Based on the one-to-one correspondence between elements of VC set and elements of POC set, the following three types of symbolic operation rules can be determined [7–11]:

(1) Operation rules for "intersection" of rotational elements [linear operation: Eqs. (5.3a)–(5.3f)].

(2) Operation rules for "intersection" of translational elements [linear operation: Eqs. (5.3g)–(5.3l)].

(3) Rules for selection of independent elements [nonlinear criteria: Eqs. (5.3m) and (5.3n)].

A. **Operation rules for "intersection" of rotational elements**

(1) $(r^1 \bigcap r^3)$:

$$\left[r^1(\| R_i)\right]_{b_i} \bigcap \left[r^3\right]_{b_j} = \left[r^1(\| R_i)\right]_{Pa} \tag{5.3a}$$

where, $r^1(\| R_i)$—there is one finite rotation in a direction parallel to axis of R_i.

(2) $(r^2 \bigcap r^3)$:

$$\left[r^2(\| \diamondsuit(R_{i_1}, R_{i_2}))\right]_{b_i} \bigcap \left[r^3\right]_{b_j} = \left[r^2(\| \diamondsuit(R_{i_1}, R_{i_2}))\right]_{Pa} \tag{5.3b}$$

where, $r^2(\| \diamondsuit(R_{i_1}, R_{i_2}))$—there are two finite rotations in any direction within a plane parallel to axis of R_{i_1} and axis of R_{i_2}, r^3—there are three finite rotations in any direction within 3D space.

(3) $(r^3 \bigcap r^3)$:

$$\left[r^3\right]_{b_i} \bigcap \left[r^3\right]_{b_j} = \left[r^3\right]_{Pa} \tag{5.3c}$$

If each branch contains an $SOC\{-\widehat{RRR}-\}$, Eq. (5.3c) can be rewritten as

$$\left[r^3(o_1)\right]_{b_i} \bigcap \left[r^3(o_2)\right]_{b_j} = \left[r^1(\| (o_1 - o_2)) \bigcup r^2\right]_{Pa}$$

where, $o_1(o_2)$—intersection of axes of three R pairs, $(o_1 - o_2)$—the line connecting points o_1 and o_2.

(4) $(r^1 \bigcap r^1)$:

$$[r^1(\| R_i)]_{b_i} \bigcap [r^1(\| R_j)]_{b_j} = \begin{cases} [r^1(\| R_i)]_{Pa}, & if\ R_i \| R_j. \\ [r^0]_{Pa}, & if\ R_i \nparallel R_j \end{cases} \quad (5.3d)$$

where, r^0—there is no finite rotation.

(5) $(r^1 \bigcap r^2)$:

$$[r^1(\| R_i)]_{b_i} \bigcap [r^2(\| \Diamond(R_{j_1}, R_{j_2}))] = \begin{cases} [r^1(\| R_i)]_{Pa}, & if\ R_i \| (\Diamond(R_{j_1}, R_{j_2})). \\ [r^0]_{Pa}, & if\ R_i \nparallel (\Diamond(R_{j_1}, R_{j_2})). \end{cases} \quad (5.3e)$$

(6) $(r^2 \bigcap r^2)$:

$$\begin{aligned} & [r^2(\| \Diamond(R_{i_1}, R_{i_2}))]_{b_i} \bigcap [r^2(\| \Diamond(R_{j_1}, R_{j_2}))]_{b_j} \\ & = \begin{cases} [r^2(\| \Diamond(R_{i_1}, R_{i_2}))]_{Pa}, & if\ (\Diamond(R_{i_1}, R_{i_2})) \| (\Diamond(R_{j_1}, R_{j_2})). \\ [r^1(\| (\Diamond(R_{i_1}, R_{i_2}) \bigcap \Diamond(R_{j_1}, R_{j_2})))]_{Pa}, & if\ (\Diamond(R_{i_1}, R_{i_2})) \nparallel (\Diamond(R_{j_1}, R_{j_2})). \end{cases} \end{aligned} \quad (5.3f)$$

where, $r^1(\| (\Diamond(R_{i_1}, R_{i_2}) \bigcap \Diamond(R_{j_1}, R_{j_2})))$—there is a finite rotation in a direction parallel to intersection line of plane $\Diamond(R_{i_1}, R_{i_2})$ and plane $\Diamond(R_{j_1}, R_{j_2})$.

B. **Operation rules for "intersection" of translational elements**

(1) $(t^1 \bigcap t^3)$:

$$[t^1(\| P_i^*)]_{b_i} \bigcap [t^3]_{b_j} = [t^1(\| P_i^*)]_{Pa} \quad (5.3g)$$

where, P^*—translation of a P pair or derivative translation of an R (H) pair, $t^1(\| P_i^*)$—there is a finite translation in a direction parallel to P^*.

It should be noted that the derivative translation of R (H) pair shall be considered when Eqs. (5.3g)–(5.3l) are used.

(2) $(t^2 \bigcap t^3)$:

$$[t^2(\| \Diamond(P_{i_1}^*, P_{i_2}^*))]_{b_i} \bigcap [t^3]_{b_j} = [t^2(\| \Diamond(P_{i_1}^*, P_{i_2}^*))]_{Pa} \quad (5.3h)$$

where, $t^2(\parallel \Diamond(P^*_{i_1}, P^*_{i_2}))$—there are two finite translations in any direction within a plane parallel to $P^*_{i_1}$ and $P^*_{i_2}$, t^3—the moving platform has three finite translations.

(3) $(t^3 \bigcap t^3)$:

$$[t^3]_{b_i} \bigcap [t^3]_{b_j} = [t^3]_{Pa} \tag{5.3i}$$

(4) $(t^1 \bigcap t^1)$:

$$[t^1(\parallel P^*_i)]_{b_i} \bigcap \left[t^1(\parallel P^*_j)\right]_{b_j} = \begin{cases} [t^1(\parallel P^*_i)]_{Pa}, & \textit{if } P^*_i | P^*_j. \\ [t^0]_{Pa}, & \textit{if } P^*_i \nparallel P^*_j. \end{cases} \tag{5.3j}$$

where, t^0—the moving platform has no translation.

(5) $(t^1 \bigcap t^2)$:

$$[t^1(\parallel P^*_i)]_{b_i} \bigcap \left[t^2(\parallel \Diamond(P^*_{j_1}, P^*_{j_2}))\right] = \begin{cases} [t^1(\parallel P^*_i)]_{Pa}, & \textit{if } P^*_i \parallel (\Diamond(P^*_{j_1}, P^*_{j_2})). \\ [t^0]_{Pa}, & \textit{if } P^*_i \nparallel (\Diamond(P^*_{j_1}, P^*_{j_2})). \end{cases} \tag{5.3k}$$

(6) $(t^2 \bigcap t^2)$:

$$\left[t^2(\parallel \Diamond(P^*_{i_1}, P^*_{i_2}))\right]_{b_i} \bigcap \left[t^2(\parallel \Diamond(P^*_{j_1}, P^*_{j_2}))\right] = \begin{cases} \left[t^2(\parallel \Diamond(P^*_{i_1}, P^*_{i_2}))\right]_{Pa}, & \textit{if}(\Diamond(P^*_{i_1}, P^*_{i_2})) \parallel (\Diamond(P^*_{j_1}, P^*_{j_2})). \\ \left[t^1(\parallel (\Diamond(P^*_{i_1}, P^*_{i_2}) \bigcap \Diamond(P^*_{j_1}, P^*_{j_2})))\right]_{Pa}, & \textit{if}(\Diamond(P^*_{i_1}, P^*_{i_2})) \nparallel (\Diamond(P^*_{j_1}, P^*_{j_2})). \end{cases} \tag{5.3l}$$

where, $t^1(\parallel (\Diamond(P^*_{i_1}, P^*_{i_2}) \bigcap \Diamond(P^*_{j_1}, P^*_{j_2})))$—there is a finite translation in a direction parallel to intersection line of plane $\Diamond(P^*_{i_1}, P^*_{i_2})$ and plane $\Diamond(P^*_{j_1}, P^*_{j_2})$.

C. **Criteria for selection of independent elements**

(1) According to Eq. (3.10) in Chap. 3, number of independent elements of the moving platform's POC set shall not be greater than DOF of the mechanism, i.e.

$$\dim\{M_{Pa}\} \le DOF \tag{5.3m}$$

where, $\dim\{M_{Pa}\}$—dimension of POC set (number of independent elements)

If DOF of the mechanism is 0, the POC set will be

$$M_{Pa} = \bigcap_{j=1}^{v+1} M_{b_j} = \begin{bmatrix} t^0 \\ r^0 \end{bmatrix}$$

where, $t^0(r^0)$—the moving platform has no translation (rotation).

(2) Any two independent elements can not be from POC set of the same R (or H) pair (5.3n)

It should be noted that these symbolic operation rules depend on only topological structure and are independent of motion position of the mechanism and it is not necessary to establish the fixed coordinate system.

5.4 Basic Characteristics of Branches

5.4.1 POC Set of Branches

According to POC equation for parallel mechanisms (Eq. 5.3), there is [7–11]

$$M_{b_i} \supseteq M_{Pa}, \quad (i = 1, 2, \ldots, v+1) \tag{5.4}$$

where, M_{b_i}—POC set of the ith branch, M_{Pa}—POC set of the parallel mechanism.

For example, if POC set of a parallel mechanism is $M_{Pa} = \begin{bmatrix} t^0 \\ r^3 \end{bmatrix}$, a branch of the parallel mechanism may have the following different POC sets:

$$M_{b_i} = \begin{bmatrix} t^0 \\ r^3 \end{bmatrix}, \begin{bmatrix} t^1 \\ r^3 \end{bmatrix}, \begin{bmatrix} t^2 \\ r^3 \end{bmatrix}, \begin{bmatrix} t^3 \\ r^3 \end{bmatrix}.$$

5.4.2 SOC Branch and HSOC Branch

(1) SOC branch

A branch which does not contain any loop(s) is called simple branch (also called SOC branch), as shown in Fig. 5.2a. POC set of the end link of an SOC branch can be determined according to POC equation for serial mechanisms (Eq. 4.4 in Chap. 4).

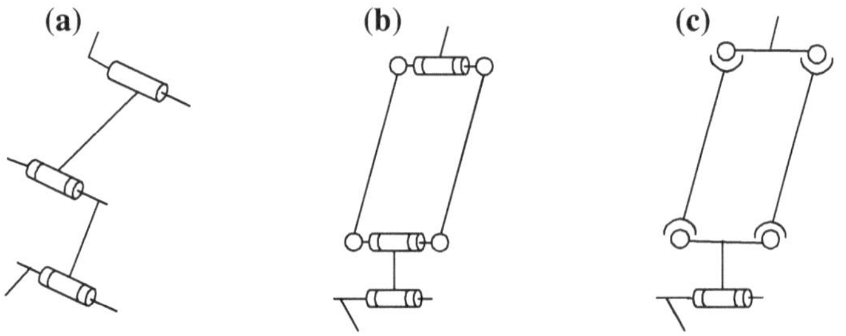

Fig. 5.2 Three topologically equivalent branches

(2) HSOC branch

A branch which contains a loop (or loops) is called complex branch (also called HSOC branch), as shown in Fig. 5.2b, c. Generally, an HSOC branch is composed of a sub-PM and several kinematic pairs connected in series. Commonly used two-branch sub-PMs are listed in Table 5.1.

For example, the HSOC in Fig. 5.2b is composed of a sub-PM with two branches (4R parallelogram) and three R pairs connected in series. The HSOC in Fig. 5.2c is composed of a sub-PM with two branches (4S parallelogram) and an R pair connected in series.

Table 5.1 Sub-PMs commonly used in HSOC branches [9–11]

No.	1	2	3	4
Sub-PMs	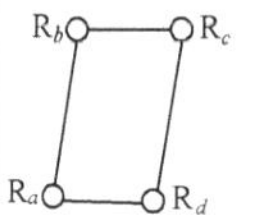	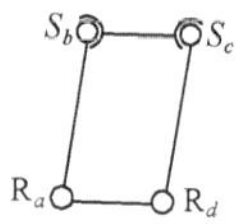	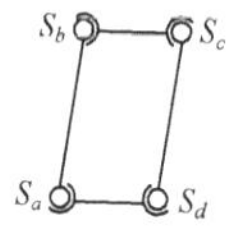	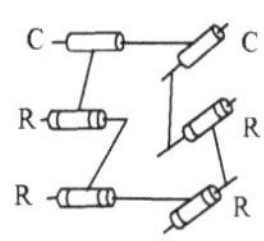
POC sets	$\begin{bmatrix} t^1(\|\|^{\perp}(ad)) \\ r^0 \end{bmatrix}$	$\begin{bmatrix} t^1(\|\|^{\perp} bc) \\ r^1(\|\| bc) \end{bmatrix}$	$\begin{bmatrix} t^1(\|\|^{\perp}(ad))\ t^1(\perp(ad)) \\ r^1(\|\|(ad)) \end{bmatrix}$	$\begin{bmatrix} t^3 \\ r^0 \end{bmatrix}$
Topologically equivalent SOCs			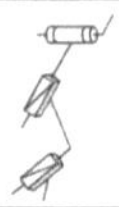	

5.4.3 Topological Equivalence Principle

(1) Topological equivalence between an SOC branch and an HSOC branch

If an SOC branch and an HSOC branch have the same POC set, the SOC branch and the HSOC branch are topologically equivalent and are called topologically equivalent branches.

For example, the SOC branch in Fig. 5.2a and the two HSOCs in Fig. 5.2b, c are topologically equivalent branches since they have the same POC set

$$M_b = \begin{bmatrix} t^3 \\ r^1(\| R) \end{bmatrix}.$$

(2) Topological equivalence between a sub-SOC and a sub-PM

If a sub-PM and a sub-SOC have the same POC set, the sub-PM and the sub-SOC are topologically equivalent.

For example, sub-$SOC\{-P-P-P-\}$ of $SOC\{-P-P-P-R-\}$ in Fig. 5.3a and the sub-PM (sub-Delta mechanism) in Fig. 5.3b are topologically equivalent branches. They have the same POC set

$$M_{sub-SOC} = M_{sub-PM} = \begin{bmatrix} t^3 \\ r^0 \end{bmatrix}$$

Since most HSOCs contain only sub-PMs with two branches, four types of such sub-PMs and their topologically equivalent sub-SOCs are listed in Table 5.1.

It should be noted that any parallel mechanism can be used as a sub-PM to replace a sub-SOC with the same POC set.

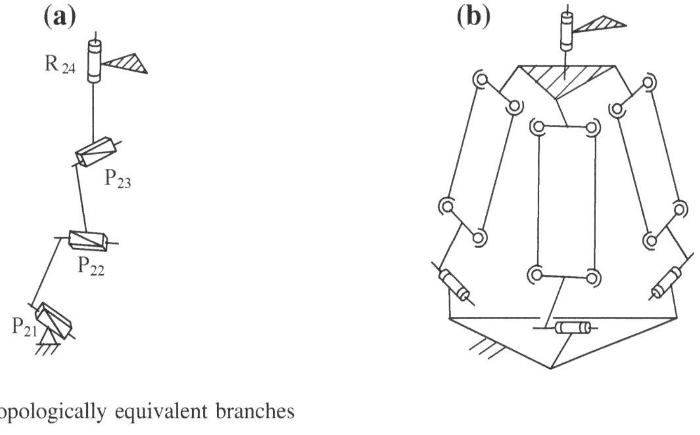

Fig. 5.3 Two topologically equivalent branches

(3) Topological equivalence principle

Generally, if the sub-PM of an HSOC branch is replaced by a topologically equivalent sub-SOC, the HSOC branch will become an SOC branch. Since the sub-PM and the sub-SOC have the same POC set, the obtained SOC branch and the original HSOC branch also have the same POC set. POC set of the parallel mechanism will also be kept unchanged. This explains the so-called topological equivalence principle.

Therefore, when sub-PM in an HSOC branch is replaced by a topologically equivalent sub-SOC, POC set of multi-loop spatial mechanisms can be calculated more easily (refer to Example 5.1).

5.5 Calculation of POC Set for Parallel Mechanisms

5.5.1 Main Steps

(1) Determine topological structure of the parallel mechanism.

(2) Select base point o′ on the moving platform (origin of the moving coordinate system).

(3) Determine POC set of each branch.

For the same base point o′ on the moving platform, determine POC set of each branch using POC equation for serial mechanisms (Eq. 4.4 in Chap. 4).

(4) Establish POC equation for parallel mechanisms.

Substitute POC set of each branch into the POC equation for parallel mechanisms (Eq. 5.3).

(5) Determine rotational elements of the moving platform's POC set based on Eqs. (5.3a)–(5.3f).

(6) Determine translational elements of the moving platform's POC set based on Eqs. (5.3g)–(5.3l).

(7) Select independent elements of the moving platform's POC set based on Eqs. (5.3m) and (5.3n). Put non-independent element in a brace.

It should be noted that calculation of POC set of a parallel mechanism always involves DOF calculation and inactive pair determination (refer to Sect. 6.3.1 and Example 6.2 in Chap. 6). But the DOF formula and the method for inactive pair determination are not introduced until Chap. 5. For all examples in the Chap. 5, DOF of the parallel mechanism is known and the mechanism contains no inactive pair. Calculation of POC set and DOF of parallel mechanism containing inactive pair shall be introduced in Chap. 6.

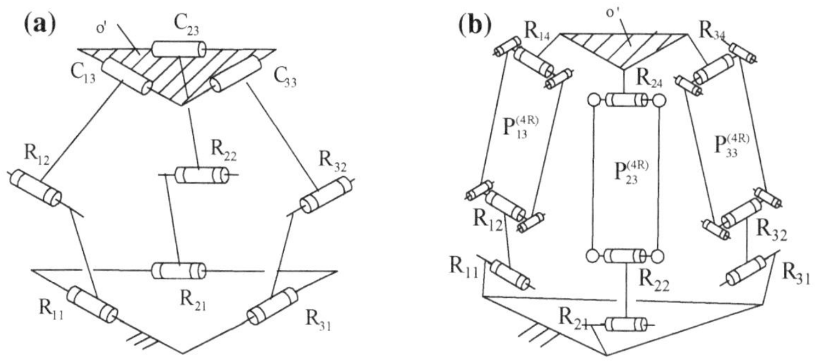

Fig. 5.4 Two 3-transloation PMs

5.5.2 Examples

Example 5.1 Determine POC sets of the two parallel mechanisms in Fig. 5.4a, b.

(1) Determine topological structure of the parallel mechanism in Fig. 5.4a

- Topological structure of branches

$$SOC\{-R_{j1} \parallel R_{j2} \parallel C_{j3}-\}, \quad \mathrm{j} = 1, 2, 3$$

- Topological structures of two platforms

Axes of R_{11}, R_{21} and R_{31} form a triangle, as shown in Fig. 5.4a.

(2) Select an arbitrary point (o') on the moving platform as base point, as shown in Fig. 5.4a.

(3) Determine POC set of each branch

For the same base point o′ on moving platform, POC set of each branch can be obtained based on POC equation for serial mechanisms (Eq. 4.4 in Chap. 4).

$$M_{bj} = \begin{bmatrix} t^3 \\ r^1(\parallel R_{j1}) \end{bmatrix}, \quad \mathrm{j} = 1,\ 2,\ 3.$$

(4) Establish POC equation for parallel mechanisms

Substitute POC set of each branch into Eq. (5.3), there is

$$\begin{aligned} M_{Pa} &= \begin{bmatrix} t^3 \\ r^1(\parallel R_{11}) \end{bmatrix} \bigcap \begin{bmatrix} t^3 \\ r^1(\parallel R_{21}) \end{bmatrix} \bigcap \begin{bmatrix} t^3 \\ r^1(\parallel R_{31}) \end{bmatrix} \\ &= \begin{bmatrix} t^3 & \cap & t^3 & \cap & t^3 \\ r^1(\parallel R_{11}) & \cap & r^1(\parallel R_{21}) & \cap & r^1(\parallel R_{31}) \end{bmatrix} \end{aligned}$$

(5) Determine rotational elements of the POC set

For the three pairs R_{11}, R_{21} and R_{31}, neither R pair is parallel to any of the other two pairs. According to operation rule Eq. (5.3d), the above POC equation can be rewritten as

$$M_{Pa} = \begin{bmatrix} t^3 & \cap & t^3 & \cap & t^3 \\ r^1(\| R_{11}) & \cap & r^1(\| R_{21}) & \cap & r^1(\| R_{31}) \end{bmatrix} = \begin{bmatrix} t^3 \cap t^3 \cap t^3 \\ r^0 \end{bmatrix}$$

(6) Determine translational elements of the POC set

According the Eq. (5.3i), the above equation can be further rewritten as

$$M_{Pa} = \begin{bmatrix} t^3 \cap t^3 \cap t^3 \\ r^0 \end{bmatrix} = \begin{bmatrix} t^3 \\ r^0 \end{bmatrix}$$

(7) Select independent elements of the POC set

Since DOF of the mechanism is three, the POC set has three independent elements according to Eq. (5.3m). So, moving platform of the parallel mechanism has three independent translations.

(8) Determine POC set of the parallel mechanism in Fig. 5.4b.

Since the HSOC branch in Fig. 5.4b and the SOC branch in Fig. 5.4a are topologically equivalent (Fig. 5.2a, b), the HSOC branch in Fig. 5.4b can be replaced by its topologically equivalent SOC branch (Fig. 5.4a). So, POC set of the parallel mechanism in Fig. 5.4b is the same as that of the parallel mechanism in Fig. 5.4a, i.e.

$$M_{Pa} = \begin{bmatrix} t^3 \\ r^0 \end{bmatrix}$$

So, moving platform of parallel mechanism in Fig. 5.4b has three independent translations.

From this example, we know that calculation of POC set may be simplied by replacing HSOC branch with topologically equivalent SOC branch.

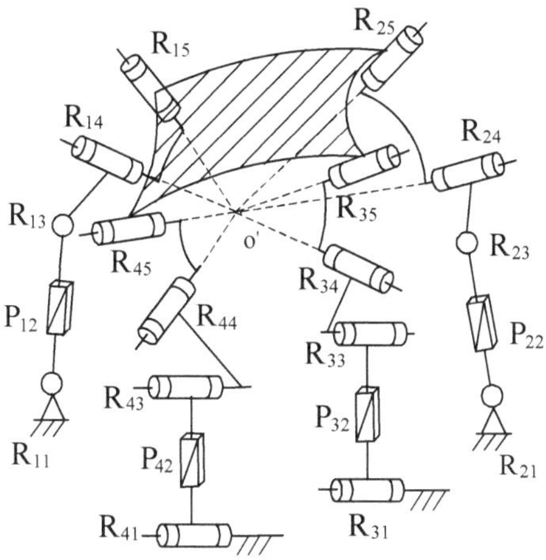

Fig. 5.5 $4 - SOC\{-R(\perp P) \parallel R - \widehat{RR} -\}$ PM

Example 5.2 Determine POC set of the parallel mechanism in Fig. 5.5.

(1) Determine topological structure of the parallel mechanism

- Topological structure of branches

$$SOC\{-R_{i1}(\perp P_{i2}) \parallel R_{i3} - \widehat{R_{i4}R_{i5}} -\}, \quad \mathrm{i} = 1, 2, 3, 4$$

- Topological structures of two platforms

  Fixed platform: $R_{11} \parallel R_{21}$; $R_{31} \parallel R_{41}$; $R_{11} \nparallel R_{31}$

  Moving platform: Axes of $R_{14}, R_{15}, R_{24}, R_{25}, R_{34}, R_{35}, R_{44}$ and R_{45} intersect at the same point o′.

(2) Select point o′ (intersection of axes of R pairs) on the moving platform as base point, as shown in Fig. 5.5.

(3) Determine POC set of each branch

According to Example 4.5 (Fig. 4.6) in Chap. 4, POC set of each branch is

$$M_{b_i} = \begin{bmatrix} t^2(\perp R_{i1}) \\ r^3 \end{bmatrix}, \quad \mathrm{i} = 1, 2, 3, 4.$$

where, $t^2(\perp R_{i1})$—the end link has two finite translations in a plane perpendicular to axis of R_{i1}.

(4) Establish POC equation for parallel mechanisms

Substitute POC set of each branch into Eq. (5.3), there is

$$M_{Pa} = \begin{bmatrix} t^2(\perp R_{11}) \\ r^3 \end{bmatrix} \bigcap \begin{bmatrix} t^2(\perp R_{21}) \\ r^3 \end{bmatrix} \bigcap \begin{bmatrix} t^2(\perp R_{31}) \\ r^3 \end{bmatrix} \bigcap \begin{bmatrix} t^2(\perp R_{41}) \\ r^3 \end{bmatrix}$$

$$= \begin{bmatrix} t^2(\perp R_{11}) \bigcap t^2(\perp R_{21}) \bigcap t^2(\perp R_{31}) \bigcap t^2(\perp R_{41}) \\ r^3 \quad \bigcap \quad r^3 \quad \bigcap \quad r^3 \quad \bigcap \quad r^3 \end{bmatrix}$$

(5) Determine rotational elements of the POC set

According to operation rule Eq. (5.3c), the above POC equation can be rewritten as

$$M_{Pa} = \begin{bmatrix} t^2(\perp R_{11}) & \bigcap & t^2(\perp R_{21}) & \bigcap & t^2(\perp R_{31}) & \bigcap & t^2(\perp R_{41}) \\ r^3 & \bigcap & r^3 & \bigcap & r^3 & \bigcap & r^3 \end{bmatrix}$$

$$= \begin{bmatrix} t^2(\perp R_{11}) \quad \bigcap \quad t^2(\perp R_{21}) \quad \bigcap \quad t^2(\perp R_{31}) \quad \bigcap \quad t^2(\perp R_{41}) \\ r^3 \end{bmatrix}$$

(6) Determine translational elements of the POC set

Since there is "$R_{11} \parallel R_{21}, R_{31} \parallel R_{41}, R_{11} \nparallel R_{31}$", according to Eq. (5.3l), the above equation can be further rewritten as

$$M_{Pa} = \begin{bmatrix} [t^2(\perp R_{11})] \bigcap [t^2(\perp R_{21})] \bigcap [t^2(\perp R_{31})] \bigcap [t^2(\perp R_{41})] \\ r^3 \end{bmatrix}$$

$$= \begin{bmatrix} t^2(\perp R_{11}) \bigcap t^2(\perp R_{31}) \\ r^3 \end{bmatrix}$$

$$= \begin{bmatrix} t^1(\perp \diamondsuit(R_{11}, R_{31})) \\ r^3 \end{bmatrix}$$

(7) Select independent elements of the POC set

Since DOF of the mechanism is four, the POC set has four independent elements according to Eq. (5.3m). So, moving platform of the parallel mechanism has three independent rotations and one independent translation along the common perpendicular of axis of R_{11} and axis of R_{31}.

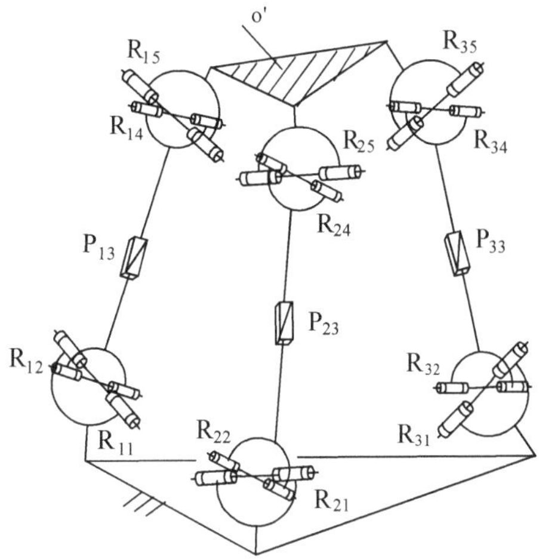

Fig. 5.6 $3-SOC\{-\widehat{R\perp R}(\perp P)\parallel\widehat{R\perp R}-\}$ PM

Example 5.3 Determine POC set of the parallel mechanism in Fig. 5.6.

(1) Determine topological structure of the parallel mechanism

- Topological structure of branches

$$SOC\{-\widehat{R_{i1}\perp R}_{i2}(\perp P_{i3})\parallel\widehat{R_{i4}\perp R_{i5}}-\},\quad i=1,2,3$$

- Topological structures of two platforms

  Fixed platform: axes of R_{11}, R_{21} and R_{31} are skew lines.

  Moving platform: ensure $R_{i1}\parallel R_{i5}$ $(\mathrm{i}=1,2,3)$ upon assembling.

(2) Select an arbitrary point o′ on the moving platform as base point.

(3) Determine POC set of each branch

According to Example 4.6 (Fig. 4.7) in Chap. 4, POC set of each branch (at the moment of assembling) is

$$M_{bi}=\begin{bmatrix} t^3 \\ r^2(\parallel\diamondsuit(R_{i4},R_{i5})) \end{bmatrix},\quad \mathrm{i}=1,2,3.$$

(4) Establish POC equation for parallel mechanisms

Substitute POC set of each branch into Eq. (5.3), there is

$$M_{Pa} = \begin{bmatrix} t^3 \\ r^2(\| \diamondsuit(R_{14}, R_{15})) \end{bmatrix} \bigcap \begin{bmatrix} t^3 \\ r^2(\| \diamondsuit(R_{24}, R_{25})) \end{bmatrix} \bigcap \begin{bmatrix} t^3 \\ r^2(\| \diamondsuit(R_{34}, R_{35})) \end{bmatrix}$$
$$= \begin{bmatrix} t^3 & \bigcap & t^3 & \bigcap & t^3 \\ r^2(\| \diamondsuit(R_{14}, R_{15}))] & \bigcap & r^2(\| \diamondsuit(R_{24}, R_{25})) & \bigcap & r^2(\| \diamondsuit(R_{34}, R_{35})) \end{bmatrix}$$

(5) Determine rotational elements of the POC set

According to operation rules Eqs. (5.3e) and (5.3f), the above POC equation can be rewritten as

$$M_{Pa} = \begin{bmatrix} t^3 & \bigcap & t^3 & \bigcap & t^3 \\ r^2(\| \diamondsuit(R_{14}, R_{15})) & \bigcap & r^2(\| \diamondsuit(R_{24}, R_{25})) & \bigcap & r^2(\| \diamondsuit(R_{34}, R_{35})) \end{bmatrix}$$
$$= \begin{bmatrix} t^3 \; \bigcap \; t^3 \; \bigcap \; t^3 \\ r^0 \end{bmatrix}$$

(6) Determine translational elements of the POC set

According to Eq. (5.3i), the above equation can be further rewritten as

$$M_{Pa} = \begin{bmatrix} t^3 \bigcap t^3 \bigcap t^3 \\ r^0 \end{bmatrix} = \begin{bmatrix} t^3 \\ r^0 \end{bmatrix}$$

(7) Select independent elements of the POC set

Since DOF of the mechanism is three, the POC set has three independent translational elements according to Eq. (5.3m). So, moving platform of the parallel mechanism has three independent translations.

(8) Discussion

"$R_{i1} \parallel R_{i5}$ (i $= 1, 2, 3$)" is a precondition for this parallel mechanism to have only three independent translations. But R_{i1} and R_{i5} are not on the same link. They are parallel just at the moment of assembling and may become unparallel during the motion process due to clearance of kinematic pair, tolerance of machining or assembling or deformation caused by external forces. This may lead to a minor rotation of the moving platform. So, this 3-translation parallel mechanism is sensitive to all kinds of errors [16].

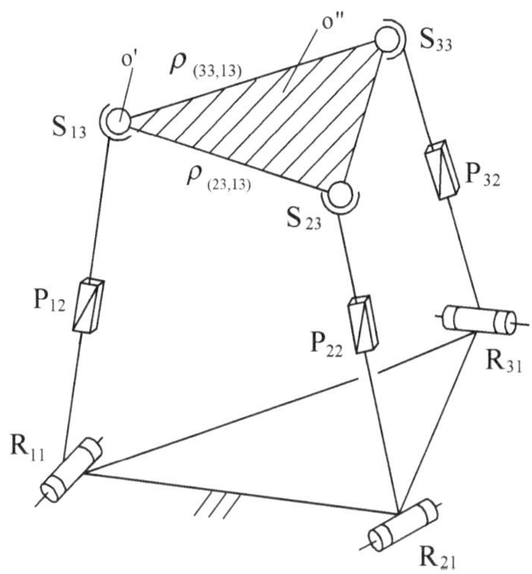

Fig. 5.7 $3 - SOC\{-R\perp P - S-\}$ PM

Example 5.4 Determine POC set of the parallel mechanism in Fig. 5.7.

(1) Determine topological structure of the parallel mechanism

- Topological structure of branches

$$SOC\{-R_{i1}\perp P_{i2} - S_{i3}-\}, \quad i = 1, 2, 3.$$

- Topological structures of two platforms

Fixed platform: Axes of the three R pairs are not parallel to a plane.

Moving platform: as shown in Fig. 5.7.

(2) Select center o_{13} of S_{13} as base point o′.

(3) Determine POC set of each branch

(a) The first branch:

This branch is composed of a sub-SOC (sub-$SOC\{-\widehat{RRR}-\}$), an R pair and a P pair connected in series. For base point o′, substititure POC set of R pair (Eq. 3.2), POC set of P pair (Eq. (3.1)) and POC set of sub-$SOC\{-\widehat{RRR}-\}$ (No. 6* in Table 4.1) into Eq. (4.4) in Chap. 4, there is

$$M_{b1} = \begin{bmatrix} t^1(\perp(R_{11})) \\ r^1(\| R_{11}) \end{bmatrix} \cup \begin{bmatrix} t^1(\| P_{12}) \\ r^0 \end{bmatrix} \cup \begin{bmatrix} t^0 \\ r^3(o_{13}) \end{bmatrix} = \begin{bmatrix} t^2(\perp R_{11}) \\ r^3(o_{13}) \end{bmatrix}$$

(b) The second branch and the third branch:

Similar to the first branch, for base point o′, substituter POC set of R pair (Eq. 3.2), POC Set of P pair (Eq. 3.1) and POC set of sub-$SOC\{-\overbrace{RRR}-\}$ (No. 6 in Table 4.1) into Eq. (4.4) in Chap. 4, there is

$$
\begin{aligned}
M_{bi} &= \begin{bmatrix} t^1(\perp R_{i1}) \\ r^1(\parallel R_{i1}) \end{bmatrix} \bigcup \begin{bmatrix} t^1(\parallel P_{i2}) \\ r^0 \end{bmatrix} \bigcup \begin{bmatrix} t^2(\perp \rho_{i3,13}) \\ r^3(o_{i3}) \end{bmatrix} \\
&= \begin{bmatrix} t^2(\perp R_{i1}) \bigcup t^1(\perp \rho_{i3,13}) \\ r^3(o_{i3}) \end{bmatrix} = \begin{bmatrix} t^3 \\ r^3(o_{i3}) \end{bmatrix}, \quad i = 2, 3.
\end{aligned}
$$

(4) Establish POC equation for parallel mechanisms

Substitute POC set of each branch into Eq. (5.3), there is

$$
\begin{aligned}
M_{Pa} &= \begin{bmatrix} t^2(\perp R_{11}) \\ r^3(o_{13}) \end{bmatrix} \bigcap \begin{bmatrix} t^3 \\ r^3(o_{23}) \end{bmatrix} \bigcap \begin{bmatrix} t^3 \\ r^3(o_{33}) \end{bmatrix} \\
&= \begin{bmatrix} t^2(\perp R_{11}) \bigcap t^3 \bigcap t^3 \\ r^3(o_{13}) \bigcap r^3(o_{23}) \bigcap r^3(o_{33}) \end{bmatrix}
\end{aligned}
$$

(5) Determine rotational elements of the POC set

Since axes of the three R pairs are not parallel to the same plane, according to operation rule Eq. (5.3c), the above POC equation can be rewritten as

$$
\begin{aligned}
M_{Pa} &= \begin{bmatrix} t^2(\perp R_{11}) \bigcap t^3 \bigcap t^3 \\ r^3(o_{13}) \bigcap r^3(o_{23}) \bigcap r^3(o_{33}) \end{bmatrix} \\
&= \begin{bmatrix} t^2(\perp R_{11}) \bigcap t^3 \bigcap t^3 \\ r^1(\parallel (o_{13} - o_{23})) \quad r^1(\parallel (o_{13} - o_{33})) \quad r^1(\nparallel \diamondsuit(o_{13}, o_{23}, o_{33})) \end{bmatrix}
\end{aligned}
$$

where, $\diamondsuit(o_{13}, o_{23}, o_{33})$—plane formed by centers of the three S pairs.

(6) Determine translational elements of the POC set

According to Eq. (5.3h), the above equation can be further rewritten as

$$
\begin{aligned}
M_{Pa} &= \begin{bmatrix} t^2(\perp R_{11}) \bigcap t^3 \bigcap t^3 \\ r^1(\parallel (o_{13} - o_{23})) \quad r^1(\parallel (o_{13} - o_{33})) \quad r^1(\nparallel \diamondsuit(o_{13}, o_{23}, o_{33})) \end{bmatrix} \\
&= \begin{bmatrix} t^2(\perp R_{11}) \\ r^1(\parallel (o_{13} - o_{23})) \quad r^1(\parallel (o_{13} - o_{33})) \quad r^1(\nparallel \diamondsuit(o_{13}, o_{23}, o_{33})) \end{bmatrix}
\end{aligned}
$$

(7) Select independent elements of the POC set

Since DOF of the mechanism is three, the POC set has three independent elements according to Eq. (5.3m). The other two elements are non-independent elements.

(8) Discussion

If axes of the three R pairs on the fixed platform are parallel to a plane, a new parallel mechanism is obtained. Follow above steps (1) to (7) and obtain POC set of the new parallel mechanism as follows

$$M_{Pa} = \begin{bmatrix} & t^2(\perp R_{11}) & \\ r^1(\| \, (o_{13} - o_{23})) & r^1(\| \, (o_{13} - o_{33})) & r^1(\nparallel \diamondsuit(o_{13}, o_{23}, o_{33})) \end{bmatrix}$$

Althrough this POC set has the same form as that of the original parallel mechanism, it should be noted that since each S pair can move only in the plane perpendicular to axis of the R pair and axes of these three R pairs are parallel to the same plane, the moving platform can not rotate around its normal line now. The moving platform has only two rotations within the plane parallel to axes of the three R pairs. So, the POC set should be rewritten as

$$M_{Pa} = \begin{bmatrix} t^2(\perp R_{11}) \\ r^2(\| \diamondsuit(R_{11}, R_{21}, R_{31})) \end{bmatrix}$$

where, $r^2(\| \diamondsuit(R_{11}, R_{21}, R_{31}))$—the moving platform has two rotations within the plane parallel to axes of R_{i1} $(i = 1, 2, 3.)$.

Since DOF of the mechanism is three, POC set of the mechanism has three independent elements. So, any three elements can be selected as the independent elements. The other one is the non-independent element.

Example 5.5 Determine POC set of the parallel mechanism in Fig. 5.8.

(1) Determine topological structure of the parallel mechanism

- Topological structure of branches

$$SOC\{-R_{i1} \| R_{i2} \| R_{i3} - \widehat{R_{i4}R_{i5}} -\}, \quad i = 1, 2, 3.$$

- Topological structures of two platforms

  Fixed platform: axes of R_{11}, R_{21} and R_{31} are skew lines.

  Moving platform: axes of R_{14}, R_{15}, R_{24}, R_{25}, R_{34} and R_{35} intersect at the same point o′, as shown in Fig. 5.8.

(2) Select point o′ on the moving platform as base point.

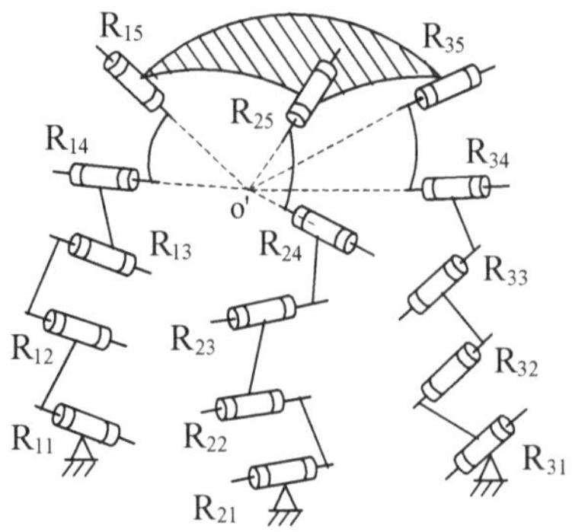

Fig. 5.8 A spherical parallel mechanism

(3) Determine POC set of each branch

According to Example 4.5 (Fig. 4.6) in Chap. 4, POC set of each branch is

$$M_{b_i} = \begin{bmatrix} t^2(\perp R_{i1}) \\ r^3 \end{bmatrix}, \quad \mathrm{i} = 1, 2, 3.$$

where, $t^2(\perp R_{i1})$—the end link has two finite translations within a plane perpendicular to axis of R_{i1}.

(4) Establish POC equation for parallel mechanisms

Substitute POC set of each branch into Eq. (5.3), there is

$$\begin{aligned} M_{Pa} &= \begin{bmatrix} t^2(\perp R_{11}) \\ r^3 \end{bmatrix} \bigcap \begin{bmatrix} t^2(\perp R_{21}) \\ r^3 \end{bmatrix} \bigcap \begin{bmatrix} t^2(\perp R_{31}) \\ r^3 \end{bmatrix} \\ &= \begin{bmatrix} t^2(\perp R_{11}) & \bigcap & t^2(\perp R_{21}) & \bigcap & t^2(\perp R_{31}) \\ r^3 & \bigcap & r^3 & \bigcap & r^3 \end{bmatrix} \end{aligned}$$

(5) Determine rotational elements of the POC set

According to operation rule Eq. (5.3c), the above POC equation can be rewritten as

$$\begin{aligned} M_{Pa} &= \begin{bmatrix} t^2(\perp R_{11}) \\ r^3 \end{bmatrix} \bigcap \begin{bmatrix} t^2(\perp R_{21}) \\ r^3 \end{bmatrix} \bigcap \begin{bmatrix} t^2(\perp R_{31}) \\ r^3 \end{bmatrix} \\ &= \begin{bmatrix} t^2(\perp R_{11}) \cap t^2(\perp R_{21}) \cap t^2(\perp R_{31}) \\ r^3 \end{bmatrix} \end{aligned}$$

(6) Determine translational elements of the POC set

Since axes of R_{11}, R_{21} and R_{31} are skew lines, according to operation rules Eqs. (5.3l) and (5.3k), the above equation can be further rewritten as

$$M_{Pa} = \begin{bmatrix} t^2(\perp R_{11}) \\ r^3 \end{bmatrix} \bigcap \begin{bmatrix} t^2(\perp R_{21}) \\ r^3 \end{bmatrix} \bigcap \begin{bmatrix} t^2(\perp R_{31}) \\ r^3 \end{bmatrix} = \begin{bmatrix} t^0 \\ r^3 \end{bmatrix}$$

(7) Select independent elements of the POC set

Since DOF of the mechanism is three, the POC set has three independent elements. So, moving platform of the parallel mechanism in Fig. 5.8 has three independent rotations.

5.6 Summary

(1) Based on topological structure invariance of parallel mechanisms (excluding singular position) and unit vector set of the moving platform velocity, VC equation for parallel mechanisms (Eq. (5.2)) and the corresponding operation rules are derived. This VC equation and the corresponding symbolic operation rules also have invariance.

(2) Based on invariance property of the moving platform's VC set, POC equation for parallel mechanisms (Eq. 5.3) and the corresponding symbolic operation rules are obtained. This POC equation and the corresponding operation rules also have invariance property. Determination of the moving platform's POC set always involves DOF calcualation and inactive pair judgment (refer to Chap. 6 for detail).

(3) Based on POC equation for parallel mechanisms, the formula for determining POC set of branches (Eq. 5.4) is derived. This formula can be used in structure synthesis of branches (refer to Chap. 9 for detail).

(4) Topological equivalence principle is proposed (Sect. 5.4.3). HSOC branches can be replaced by their topologically equivalent SOC branches in order to simplify POC set calculation of parallel mechanisms.

(5) POC equation for parallel mechanisms and the corresponding operation rules are independent of motion position of the mechanism and the fixed coordinate system (i.e. it is not necessary to establish the fixed coordinate system). So, it could be called as a geometrical method

(6) POC equation for parallel mechanisms can be used in establishment of DOF formula (Chap. 6) and topology design of parallel mechanisms (Chap. 9).

References

1. Huang Z, Li QC (2002) General methodology for type synthesis of symmetrical lower-mobility parallel manipulators and several novel manipulators. Int J Robot Res 21:131–145
2. Meng X, Gao F, Shengfu WuS, Ge QJ (2014) Type synthesis of parallel robotic mechanisms: framework and brief review. Mech Mach Theor 78:177–186
3. Herve JM (1978) Analyse structurelle des mécanismes par groupe des déplacements. Mech Mach Theor 13:437–450
4. Fanghella P, Galletti C (1995) Metric relations and displacement groups in mechanism and robot kinematics. ASME J Mech Des 117:470–478
5. Meng J, Liu GF, Li ZX (2007) A geometric theory for analysis and synthesis of Sub-6 DOF parallel manipulators. IEEE Trans Rob 23:625–649
6. Gogu G (2007) Structural synthesis of parallel robots: Part 1: methodology. Springer, Dordrecht
7. Yang T-L, Jin Q et al (2001) A general method for type synthesis of rank-deficient parallel robot mechanisms based on SOC unit. Mach Sci Technol 20(3):321–325
8. Yang T-L, Jin Q, Liu A-X, Shen H-P, Luo Y-F (2002) Structural synthesis and classification of the 3-DOF translation parallel robot mechanisms based on the unites of single-open chain. Chin J Mech Eng 38(8):31–36. (doi:10.3901/JME.2002.08.031)
9. Jin Q, Yang T-L (2004) Theory for topology synthesis of parallel manipulators and its application to three-dimension-translation parallel manipulators. ASME J Mech Des 126:625–639
10. Yang T-L (2004) Theory of topological structure for robot mechanisms. China Machine press, Beijing
11. Yang T- L, Liu A-X, Luo Y-F, Shen H-P et al (2009) Position and orientation characteristic equation for topological design of robot mechanisms. ASME J Mech Des 131:021001-1 ~ 17
12. Yang T-L, Sun D-J (2012) A general DOF formula for parallel mechanisms and multi-loop spatial mechanisms. ASME J Mech Robot 4(1):011001-1 ~ 17
13. Yang T-L, Liu A-X, Shen H-P, Luo Y-F et al (2013) On the correctness and strictness of the position and orientation characteristic equation for topological design of robot mechanisms. ASME J Mech Robot 5: 021009-1 ~ 18
14. Yang T-L, Liu A-X, Shen H-P et al (2015) Composition principle based on SOC unit and coupling degree of BKC for general spatial mechanisms. The 14th IFToMM World Congress, Taipei, Taiwan, 25–30 Oct 2015. doi Number: 10.6567/IFToMM.14TH.WC.OS13.135
15. Yang T-L, Liu A-X, Shen H-P et al (2012) Theory and application of robot mechanism topology. China Science press, Beijing
16. Han C, Kin J et al (2002) Kinematic sensitivity analysis of the 3-UPU parallel manipulators. Mech Mach Theor 37:787–798

Chapter 6
General Formula of Degrees of Freedom for Spatial Mechanisms

6.1 Introduction

Degree of freedom (DOF) is one of the basic concepts or parameters of a mechanism. Over the past 100 years, dozens of DOF formulas were proposed by different researchers [1]. Several new DOF formulas were proposed during the past decade. These DOF formulas revealed the mapping relations among topological structure, kinematic characteristics and DOF of the mechanism [2–14]. These DOF formulas and their features are briefly introduced as follows.

6.1.1 DOF Formula Based on Screw Theory

(1) DOF formula proposed by Zhao et al. [2]

$$F = 6 - d + \sum_{j=1}^{(v+1)} r_j \tag{6.1}$$

where, F—DOF of the mechanism, d—rank of the movable platform's motion-screw system, r_j—difference between number of elements and the rank of the jth branch's motion-screw system, v—number of independent loops ($v = m - n + 1$, m—number of kinematic pairs, n—number of links).

The main features of this formula as follows:

(a) The formula is based on the motion-screw system and its constraint-screw system.

(b) The result is instantaneous DOF. The full-cycle DOF verification needs to combine with the singularity analysis.

(c) Mobility calculation has considered the selection of active joints.

T.-L. Yang et al., *Topology Design of Robot Mechanisms*,
https://doi.org/10.1007/978-981-10-5532-4_6

(2) DOF formula proposed by Dai et al. [3, 4]

$$F = 6(n - m - 1) + \sum_{i=1}^{m} f_i + card(\langle S^r \rangle) - card(\{S^r\}) \tag{6.2}$$

where, f_i—DOF of the ith kinematic pair, $card(\langle S^r \rangle)$—dimension of the movable platform's motion-screw system, $card(\{S^r\})$—dimension of the maximum linearly independent group of the movable platform's motion-screw system. Other symbols are same as those in Eq. (6.1).

The main features of this formula as follows:

(a) Mobility calculation is base on the motion-screw system and its constraint-screw system.

(b) The result is instantaneous DOF. So it is necessary full-cycle DOF verification.

(3) DOF formula proposed by Kong and Gosselin [5, 6]

$$F = \sum_{i=1}^{m} f_i - \sum_{j=1}^{k} C^j + C \tag{6.3}$$

where, k—number of branches of the parallel mechanism, C^j—rank of the jth branch's motion-screw system, C—rank of the movable platform's motion-screw system. Other symbols are same as those in Eq. (6.1).

The main features of this formula as follows:

(a) Mobility calculation is based on the motion-screw system and its constraint-screw system.

(b) The result is instantaneous DOF. So it is necessary full-cycle DOF verification.

(4) DOF formula proposed by Huang et al. [7, 8]

$$F = \sum_{i=1}^{m} f_i - d(m - n + 1) + \left(\sum_{j=1}^{p} q_j - \lambda \cdot p - k \right) \tag{6.4}$$

where, F—DOF of the mechanism, d—Rank of the screw system ($d = 6 - \lambda$, λ—number of common constraints of the mechanism), p—number of branches of the parallel mechanism, q_j—number of constraint screw of each branch, k—rank of $p(q_j - \lambda)$ constraint screws.

The main features of this formula as follows:

(a) Mobility calculation is based on the motion-screw system and its constraint-screw system.

(b) The result is instantaneous DOF. So it is necessary full-cycle DOF verification.

Equations (6.1)–(6.4) have different forms, but it has been proved that they can be expressed in the following unified form [15]:

$$F = \sum_{i=1}^{m} f_i - \sum_{j=1}^{(v+1)} rank(S_{b_j}) + rank(S_{Pa}) \tag{6.5}$$

where, F—DOF of the mechanism, $rank(S_{b_j})$—rank of the jth branch's motion-screw system, $rank(S_{Pa})$—rank of the movable platform's motion-screw system. Other symbols are same as those in Eq. (6.2).

Main features of the DOF formula based on screw theory (Eq. 6.5) are:

(a) be used for parallel mechanism.

(b) instantaneous DOF is obtained. Verification of full-cycle DOF is necessary and several methods for full-cycle DOF judgment are provided [2–8].

6.1.2 DOF Formula Based on Linear Transformation

DOF formula proposed by Gogu [9]

$$F = \sum_{i=1}^{m} f_i - \sum_{j=1}^{k} \dim(R_{Aj}) + \dim\left(\bigcap_{i=1}^{k} R_{Aj}\right) \tag{6.6}$$

where, k—number of branches of the parallel mechanism, $\dim(R_{Aj})$—dimension of the relative velocity space between two end links of the jth branch. Other symbols are same as those in Eq. (6.1).

The main features of this formula as follows:

(1) Mobility calculation is based on velocity space analysis.

(2) The result is instantaneous DOF. So it is necessary full-cycle DOF verification.

(3) When third term in the right side of Eq. (6.6) contains non-independent elements, such as the PM in Fig. 6.5, it is necessary to determine number of its independent elements.

6.1.3 DOF Formula Based on Displacement Subgroup/Submanifold

DOF formula proposed by Rico et al. [10]

$$F = \sum_{i=1}^{m} f_i - \sum_{j=1}^{(v+1)} \dim(A_j^{mlf}) + \dim\left(\bigcap_{j=1}^{(v+1)} A_j^{mlf}\right) \tag{6.7}$$

where, A_j^{mlf}—displacement subgroup (submanifold) of the movable platform relative to the fixed platform when the two platforms are connected only by the jth branch, $\dim(A_j^{mlf})$—dimension of the displacement subgroup (submanifold) A_j^{mlf} (i.e. number of independent elements), k—number of branches of the parallel mechanism. Other symbols are same as those in Eq. (6.2).

The main features of this formula as follows:

(1) Mobility calculation is based on union and intersection operation of displacement subgroup.

(2) The result is full-cycle DOF for displacement subgroup mechanisms.

(3) When third term in the right side of Eq. (6.7) contains non-independent elements, such as the PM in Fig. 6.5, it is necessary to determine number of its independent elements.

6.1.4 DOF Formula Based on POC Set

DOF formula proposed by Yang and Sun [11–14]

$$\begin{cases} F = \sum_{i=1}^{m} f_i - \sum_{j=1}^{v} \xi_{L_j} \\ \xi_{L_j} = \dim\left\{ \left(\bigcap_{i=1}^{j} M_{b_i}\right) \bigcup M_{b_{(j+1)}} \right\} \end{cases} \tag{6.8a, b}$$

This DOF formula and its main characteristics will be introduced in detail in Sect. 6.2.

6.2 The General DOF Formula

A parallel mechanism with v independent loops can be regarded as being formed by a fixed platform, a movable platform and $(v+1)$ branches connected in parallel between the fixed platform and the movable platform: the first two branches form the first independent loop (regarded as a sub-PM), the sub-PM containing the first two branches and the third branch form the second independent loop, …, the sub-PM containing the first j branches and the (j + 1)th branch form the jth independent loop, …, the sub-PM containing the first v branches and the $(v+1)$th branch form the vth independent loop, as shown in Fig. 6.1. DOF of the parallel mechanism is [11–14]

$$\begin{cases} F = \sum_{i=1}^{m} f_i - \sum_{j=1}^{v} \xi_{L_j} \\ \xi_{L_j} = \dim\left\{\left(\bigcap_{i=1}^{j} M_{b_i}\right) \bigcup M_{b_{(j+1)}}\right\} \end{cases} \tag{6.8a, b}$$

or

$$F = \sum_{i=1}^{m} f_i - \sum_{j=1}^{v} \dim\left\{\left(\bigcap_{i=1}^{j} M_{b_i}\right) \bigcup M_{b(j+1)}\right\}$$

where, F—DOF of the mechanism, f_i—DOF of the ith kinematic pair, m—number of kinematic pairs, v—number of independent loops ($v = m - n + 1$, n—number of links), ξ_{L_j}—number of independent displacement equations of the jth independent loop (formed by the topologically equivalent SOC of the a sub-PM formed by the first j branches and the (j + 1)th branch in Fig. 6.1), M_{b_i}—POC set of the ith branch's end link (Eq. (4.4) in Chap. 4), $\bigcap_{i=1}^{j} M_{b_i}$—POC set of the sub-PM formed by the first j branches (Eq. (5.3) in Chap. 5), $M_{b_{(j+1)}}$—POC set of the (j + 1)th branch's end link (Eq. 4.4).

Equation (6.8) reveals the internal relations among mechanism topological structure, DOF and POC set.

Its main features of Eq. (6.8) as follows:

(1) DOF calculation is based on POC equation of serial mechanisms and POC equation of PMs.

(2) Full-cycle DOF is obtained since this formula is independent of motion position of mechanism and the fixed coordinate system.

(3) It can be used for a mechanism whose POC set contains non-independent elements (such as the PM in Fig. 6.5).

(4) It can be used for PMs and multi-loop spatial mechanisms.

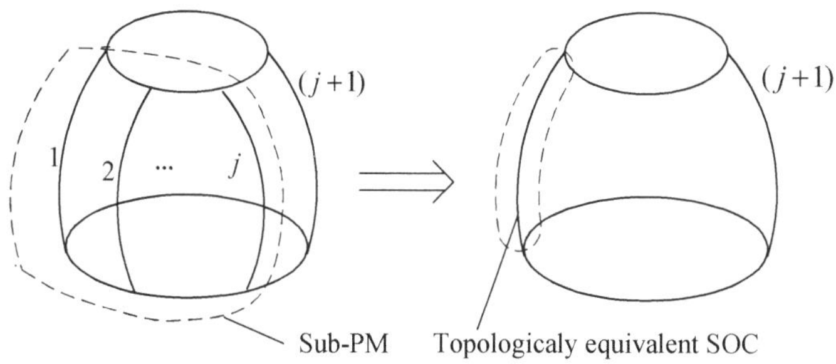

Fig. 6.1 The jth independent loop of a PM

According to set theory, Eq. 6.8 can be expressed in the following form

$$F = \sum_{i=1}^{m} f_i - \sum_{k=1}^{(v+1)} \dim\{M_{b_k}\} + \dim\left\{\bigcap\nolimits_{k=1}^{(v+1)} M_{b_k}\right\} \tag{6.9}$$

Proof

For set A and set B, dimension of $A \cup B$ is

$$\dim\{A \cup B\} = \dim\{A\} + \dim\{B\} - \dim\{A \cap B\}$$

Substitute the above equation into Eq. (6.8b), there is

$$\xi_{L_1} = \dim\{M_{b_1} \cup M_{b_2}\} = \dim\{M_{b_1}\} + \dim\{M_{b_2}\} - \dim\{M_{b_1} \cap M_{b_2}\}$$

$$\begin{aligned}\xi_{L_2} &= \dim\{(M_{b_1} \cap M_{b_2}) \cup M_{b_3}\}\\ &= \dim\{M_{b_1} \cap M_{b_2}\} + \dim\{M_{b_3}\} - \dim\{M_{b_1} \cap M_{b_2} \cap M_{b_3}\}\end{aligned}$$

....

$$\begin{aligned}\xi_{L_{(j-1)}} &= \dim\left\{\left(\bigcap\nolimits_{k=1}^{j-1} M_{b_j}\right) \cup M_{b_j}\right\}\\ &= \dim\left\{\bigcap\nolimits_{k=1}^{j-1} M_{b_j}\right\} + \dim\{M_{b_j}\} - \dim\left\{\bigcap\nolimits_{k=1}^{j} M_{b_j}\right\}\end{aligned}$$

$$\begin{aligned}\xi_{L_j} &= \dim\left\{\left(\bigcap\nolimits_{k=1}^{j} M_{b_j}\right) \cup M_{b_{(j+1)}}\right\}\\ &= \dim\left\{\bigcap\nolimits_{k=1}^{j} M_{b_j}\right\} + \dim\{M_{b_{(j+1)}}\} - \dim\left\{\bigcap\nolimits_{k=1}^{j+1} M_{b_j}\right\}\end{aligned}$$

Substitute the above results into Eqs. (6.8a), (6.9) is obtained.

In Eq. (6.9), the item "$\dim\left\{\bigcap\nolimits_{k=1}^{(v+1)} M_{b_k}\right\}$" is dimension of the POC set of the movable platform. It should not be greater than DOF of the mechanism (Eq. (3.10) in Chap. 3).

6.3 Applications of the General DOF Formula

The general DOF formula can certainly be used to calculate DOF of parallel mechanisms and multi-loop spatial mechanisms. In addition, it can also be used in the following applications.

6.3.1 Criteria for Determination of Inactive Pairs

(1) Inactive kinematic pair

Definition: for a mechanism whose DOF is greater than 0, if two links connected by a certain kinematic pair have no relative motion (or a multi-DOF kinematic pair loses one DOF), this kinematic pair is called inactive kinematic pair (or inactive DOF of a multi-DOF kinematic pair).

(2) Criteria for inactive pair determination [14]

For a mechanism whose DOF is greater than zero, supposing a certain kinematic pair (or a certain DOF of a multi-DOF kinematic pair) is locked, a new mechanism will be obtained. If DOF of the new mechanism is same as DOF of the original mechanism, the locked pair (or DOF) is an inactive pair of the original mechanism (or an inactive DOF of the multi-DOF pair). Otherwise, it is the active joint.

6.3.2 Criteria for Selection of Driving Pairs

Generally, all driving pairs should be allocated on the same platform for parallel mechanisms. Selection of driving pairs should follow the follow criteria [14]:

For a mechanism whose DOF is F, F kinematic pairs are preliminarily selected as the driving pairs. Supposing these F preliminary driving pairs are locked, a new mechanism is obtained. If DOF of this new mechanism is zero, these F kinematic pairs can be selected as the driving pairs simultaneously. If DOF of this new mechanism is greater than zero, these F kinematic pairs cannot be selected as the driving pairs simultaneously.

6.3.3 Number of Independent Displacement Equations

The second item at right side of Eq. (6.8a) is number of independent displacement equation of the mechanism. Substitute Eq. (6.8b) into this item and obtain number of independent displacement equations of the mechanism as follows [15]

$$\xi_{mech} = \sum_{j=1}^{v} \left(\dim\left\{ \left(\bigcap_{i=1}^{j} M_{b_i} \right) \bigcup M_{b_{(j+1)}} \right\} \right) \tag{6.10}$$

where, ξ_{mech}—number of independent displacement equations of the mechanism. Other symbols are same as those in Eq. (6.8).

6.3.4 Number of Overconstraints

Number of overconstraints of the mechanism is [15]

$$N_{ov.} = 6v - \xi_{mech} \tag{6.11}$$

where, $N_{ov.}$—number of overconstraints of the mechanism, v—number of independent loops, ξ_{mech}—number of independent displacement equations.

6.3.5 Redundancy

Definition: Redundancy of the mechanism is defined as [15]

$$D_{red.} = F - \dim\{M_{Pa}\} \tag{6.12a}$$

where, D_{Red}—redundancy of the mechanism, M_{Pa}—POC set of the movable platform, F–DOF of the mechanism.

Substitute Eq. (6.9) into Eq. (6.12a), there isv

$$D_{red} = \sum_{i=1}^{m} f_i - \sum_{k=1}^{(v+1)} \dim\{M_{b_k}\} \tag{6.12b}$$

According to Eq. (6.12b), redundancy is the sum of each branch's passive DOF when the movable platform is fixed.

6.4 DOF Calculation

6.4.1 Main Steps

(1) Symbolic representation of the mechanism topological structure.

(2) Select base point o' on the movable platform (origin of the moving coordinate system). The base point should be so selected that the number of non-independent element in the POC set shall be as few as possible (refer to Sect. 4.6).

(3) Determine POC set of each branch according to Eq. (4.4) in Chap. 4.

(4) Determine number of independent displacement equations according to Eq. (6.8b).

(5) Determine DOF of the PM according to Eq. (6.8a).

(6) Determine inactive pair according to criteria for inactive pair determination (Sect. 6.3.1).

(7) Determine POC set of the PM according to Eq. (5.3) in Chap. 5.

(8) Select driving pairs according to criteria for driving pair selection (Sect. 6.3.2).

For parallel mechanisms or multi-loop spatial mechanisms containing sub-PMs, replacing the sub-PM with topologically equivalent sub-SOC may simplify DOF calculation (refer to Examples 6.5 and 6.6).

6.4.2 Examples

Example 6.1 Determine DOF of the parallel mechanism in Fig. 6.2a and whether or not the three translations of C_{13}, C_{23} and C_{33} pairs can be selected as driving pairs simultaneously.

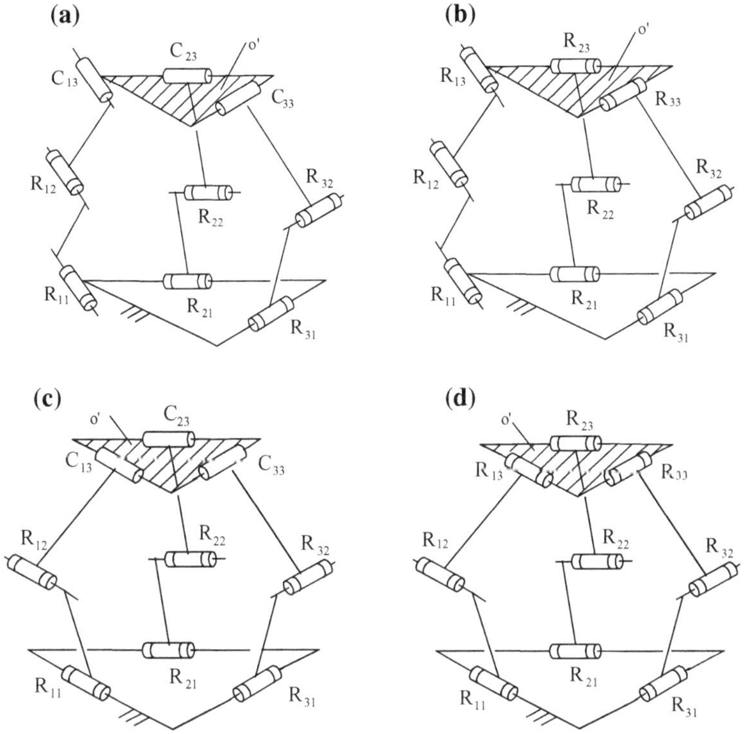

Fig. 6.2 $3 - SOC\{-R \parallel R \parallel C-\}$PM

(1) Topological structure

- Topological structure of branches:

  $SOC_i\{-R_{i1} \parallel R_{i2} \parallel C_{i3}-\}, i = 1, 2, 3$

  or $SOC_i\{-R_{i1} \parallel R_{i2} \parallel R_{i3}|P_{i3}-\}, i = 1, 2, 3.$

- Topological structure of two platforms

  Fixed platform: R_{11}, R_{21} and R_{31} are skew in space.

  Movable platform: C_{13}, C_{23} and C_{33} are skew in space.

(2) Select an arbitrary point o' on the movable platform as the base point.

(3) Determine POC set of branches.

Substituting the POC set of $SOC\{-R \parallel R \parallel R-\}$ in Table 4.1 in Chap. 4 and the POC set of P pair (Eq. (3.1)) into Eq. (4.4) in Chap. 4, POC set of each branch can be obtained as

$$M_{b_i} = \begin{bmatrix} t^3 \\ r^1(\| R_{i1}) \end{bmatrix}, \quad i = 1, 2, 3.$$

(4) Determine number of independent displacement equations of each independent loop

(a) For topological structure of the first independent loop and Eq. (6.8b), there is

$$\begin{aligned} \xi_{L_1} &= \dim\{M_{b_1} \cup M_{b_2}\} = \dim\left\{\begin{bmatrix} t^3 \\ r^1(\| R_{11}) \end{bmatrix} \cup \begin{bmatrix} t^3 \\ r^1(\| R_{21}) \end{bmatrix}\right\} \\ &= \dim\left\{\begin{bmatrix} t^3 \\ r^2(\| \diamondsuit(R_{11}, R_{21})) \end{bmatrix}\right\} = 5 \end{aligned}$$

(b) For topological structure of the second independent loop and Eq. (6.8b), there is

$$\begin{aligned} \xi_{L_2} &= \dim\{(M_{b_1} \cap M_{b_2}) \cup M_{b_3}\} \\ &= \dim\left\{\left(\begin{bmatrix} t^3 \\ r^1(\| R_{11}) \end{bmatrix} \cap \begin{bmatrix} t^3 \\ r^1(\| R_{21}) \end{bmatrix}\right) \cup \begin{bmatrix} t^3 \\ r^1(\| R_{31}) \end{bmatrix}\right\} \\ &= \dim\left\{\begin{bmatrix} t^3 \\ r^0 \end{bmatrix} \cup \begin{bmatrix} t^3 \\ r^1(\| R_{31}) \end{bmatrix}\right\} = 4. \end{aligned}$$

(5) Calculate DOF of the mechanism

According to Eq. (6.8a), DOF of the parallel mechanism is

$$F = \sum_{i=1}^{m} f_i - \sum_{j=1}^{2} \xi_{L_j} = 12 - (5 + 4) = 3$$

(6) Determine inactive pairs

In order to determine whether R_{31} (take R_{31} as an example) is an inactive pair, lock it and a new mechanism is obtained whose third branch is $SOC_3\{-R_{32} \| C_{33}-\}$. According to Eq. (4.4) in Chap. 4, POC set of the third branch is

$$M_{b_3} = \begin{bmatrix} \{t^1(\perp R_{32})\} \cup t^1(\perp R_{33}) \cup t^1(\| P_{33}) \\ r^1(\| R_{32}) \end{bmatrix} = \begin{bmatrix} \{t^1\} \cup t^2 \\ r^1(\| R_{32}) \end{bmatrix}$$

In this equation, POC set of R_{32} has only one independent element (Table 3.1 in Chap. 3). The non-independent element is put into a brace.

Now, let us determine DOF of the new mechanism.

(a) Determine ξ_{L_1} of the first independent loop
According to topological structure of the first independent loop and Eq. (6.8b), there is

$$\xi_{L_1} = \dim\{M_{b_1} \cup M_{b_2}\} = \dim\left\{\begin{bmatrix} t^3 \\ r^1(\| R_{11}) \end{bmatrix} \cup \begin{bmatrix} t^3 \\ r^1(\| R_{21}) \end{bmatrix}\right\}$$
$$= \dim\left\{\begin{bmatrix} t^3 \\ r^2(\| \diamondsuit(R_{11}, R_{21})) \end{bmatrix}\right\} = 5$$

(b) Determine ξ_{L_2} of the second independent loop
According to topological structure of the second independent loop and Eq. (6.8b), there is

$$\xi_{L_2} = \dim\{(M_{b_1} \cap M_{b_2}) \cup M_{b_3}\}$$
$$= \dim\left\{\left(\begin{bmatrix} t^3 \\ r^1(\| R_{11}) \end{bmatrix} \bigcap \begin{bmatrix} t^3 \\ r^1(\| R_{21}) \end{bmatrix}\right) \bigcup \begin{bmatrix} \{t^1\} \cup t^2 \\ r^1(\| R_{32}) \end{bmatrix}\right\}$$
$$= \dim\left\{\begin{bmatrix} t^3 \\ r^0 \end{bmatrix} \bigcup \begin{bmatrix} \{t^1\} \cup t^2 \\ r^1(\| R_{32}) \end{bmatrix}\right\} = \dim\left\{\begin{bmatrix} t^3 \\ r^1(\| R_{32}) \end{bmatrix}\right\} - 4.$$

(c) Calculate DOF of the new mechanism
According to Eq. (6.8a), DOF of the new mechanism is

$$F^* = \sum_{i=1}^{m} f_i - \sum_{j=1}^{2} \xi_j = (4 + 4 + 3) - (5 + 4) = 2$$

Since DOF of the new mechanism is not equal to DOF of the original mechanism ($F^* \neq F$), according to the criterion for determination of inactive pairs in the Sect. 6.3.1, R_{31} is not an inactive pair. Similarly, it is easy to prove that this PM has no inactive pair.

(7) Determine POC set of the parallel mechanism

According to Eq. (5.3) in Chap. 5, POC set of the parallel mechanism is

$$M_{Pa} = M_{b_1} \cap M_{b_2} \cap M_{b_3}$$
$$= \begin{bmatrix} t^3 \\ r^1(\| R_{11}) \end{bmatrix} \cap \begin{bmatrix} t^3 \\ r^1(\| R_{21}) \end{bmatrix} \cap \begin{bmatrix} t^3 \\ r^1(\| R_{31}) \end{bmatrix} = \begin{bmatrix} t^3 \\ r^0 \end{bmatrix}$$

Since DOF of the parallel mechanism is three, movable platform of this parallel mechanism has only three translations.

(8) Select driving pairs

If three translations of C_{13}, C_{23} and C_{33} pairs are selected as driving pairs and these three translations are locked, a new mechanism is obtained as shown in Fig. 6.2b.

According to Eq. (4.4) in Chap. 4, POC set of each new branch is

$$M_{b_i} = \begin{bmatrix} t^2(\perp R_{i1}) \\ r^1(\| R_{i1}) \end{bmatrix}, \ i = 1, 2, 3.$$

Now, let us determine DOF of this new mechanism.

(a) Determine number of independent displacement equations of each independent loop
According to topological structure and Eq. (6.8b), there are

$$\begin{aligned} \xi_{L_1} &= \dim\{M_{b_1} \cup M_{b_2}\} = \dim\left\{\begin{bmatrix} t^2(\perp R_{11}) \\ r^1(\| R_{11}) \end{bmatrix} \cup \begin{bmatrix} t^2(\perp R_{21}) \\ r^1(\| R_{21}) \end{bmatrix}\right\} \\ &= \dim\left\{\begin{bmatrix} t^3 \\ r^2(\| \diamondsuit(R_{11}, R_{21})) \end{bmatrix}\right\} = 5 \end{aligned}$$

$$\begin{aligned} \xi_{L_2} &= \dim\{(M_{b_1} \cap M_{b_2}) \cup M_{b_3}\} \\ &= \dim\left\{\left(\begin{bmatrix} t^2(\perp R_{11}) \\ r^1(\| R_{11}) \end{bmatrix} \cap \begin{bmatrix} t^2(\perp R_{21}) \\ r^1(\| R_{21}) \end{bmatrix}\right) \cup \begin{bmatrix} t^2(\perp R_{31}) \\ r^1(\| R_{31}) \end{bmatrix}\right\} \\ &= \dim\left\{\left(\begin{bmatrix} t^1(\perp(R_{11}, R_{21})) \\ r^0 \end{bmatrix}\right) \cup \begin{bmatrix} t^2(\perp R_{31}) \\ r^1(\| R_{31}) \end{bmatrix}\right\} \\ &= \dim\left\{\begin{bmatrix} t^3 \\ r^1(\| R_{31}) \end{bmatrix}\right\} = 4 \end{aligned}$$

(b) Determine DOF of the new mechanism
According to Eq. (6.8a), DOF of the new mechanism is

$$F^* = \sum_{i=1}^{m} f_i - \sum_{j=1}^{2} \xi_{L_j} = 9 - (5 + 4) = 0$$

Since $F^* = 0$, according to the criterion for selection of driving pairs in the Sect. 6.3.2, three translations of C_{13}, C_{23} and C_{33} pairs in Fig. 6.2a can be used as driving pairs simultaneously.

(9) Discussion

For the PM in Fig. 6.2c in which the R_{11}, R_{21} and R_{31} are in the same plane, if three translations of C_{13}, C_{23} and C_{33} pairs in Fig. 6.2c are selected as driving pairs and these three translations are locked, then, the new mechanism has been obtained as shown in Fig. 6.2d.

According to Eq. (4.4) in Chap. 4, POC of each new branch is

$$M_{b_i} = \begin{bmatrix} t^2 \\ r^1(\| R_{i1}) \end{bmatrix}, \quad i = 1, 2, 3.$$

Now, let us determine DOF of this new mechanism.

(a) Determine number of independent displacement equations of each independent loop
According to topological structure (three R pairs are in the same plane) and Eq. (6.8b), there are

$$\begin{aligned} \xi_{L_1} &= \dim\{M_{b_1} \cup M_{b_2}\} = \dim\left\{\begin{bmatrix} t^2(\perp R_{11}) \\ r^1(\| R_{11}) \end{bmatrix} \cup \begin{bmatrix} t^2(\perp R_{21}) \\ r^1(\| R_{21}) \end{bmatrix}\right\} \\ &= \dim\left\{\begin{bmatrix} t^3 \\ r^2(\| \diamondsuit(R_{11}, R_{21})) \end{bmatrix}\right\} = 5 \end{aligned}$$

$$\begin{aligned} \xi_{L_2} &= \dim\{(M_{b_1} \cap M_{b_2}) \cup M_{b_3}\} \\ &= \dim\left\{\left(\begin{bmatrix} t^2(\perp R_{11}) \\ r^1(\| R_{11}) \end{bmatrix} \cap \begin{bmatrix} t^2(\perp R_{21}) \\ r^1(\| R_{21}) \end{bmatrix}\right) \cup \begin{bmatrix} t^2(\perp R_{31}) \\ r^1(\| R_{31}) \end{bmatrix}\right\} \\ &= \dim\left\{\begin{bmatrix} t^1(\perp(R_{11}, R_{21})) \\ r^0 \end{bmatrix} \cup \begin{bmatrix} t^2(\perp R_{31}) \\ r^1(\| R_{31}) \end{bmatrix}\right\} \\ &= \dim\left\{\begin{bmatrix} t^2(\perp R_{31}) \\ r^1(\| R_{31}) \end{bmatrix}\right\} = 3 \end{aligned}$$

(b) Determine DOF of the new mechanism
According to Eq. (6.8a), DOF of the new mechanism is

$$F^* = \sum_{i=1}^{m} f_i - \sum_{j=1}^{2} \xi_{L_j} = 9 - (5 + 3) = 1$$

Since $F^* = 1 > 0$, according to the criterion for selection of driving pairs in the Sect. 6.3.2, three translations of C_{13}, C_{23} and C_{33} pairs in Fig. 6.2c can not be used as driving pairs simultaneously.

It should be noted that this result is different from that of the mechanism in Fig. 6.2a.

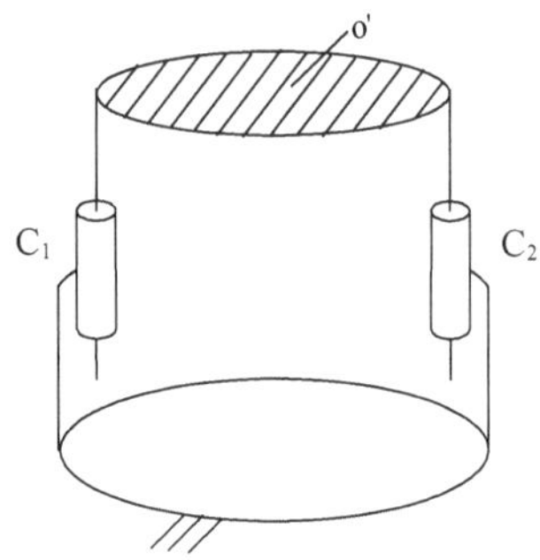

Fig. 6.3 $2-SOC\{-C-\}$PM

Example 6.2 Determine DOF of the parallel mechanism in Fig. 6.3 and POC set of its movable platform.

(1) Topological structure

- Topological structure of branches:
  $SOC\{-C_i-\}$ (or $SOC\{-R_i|P_i-\}$), i = 1,2
- Topological structure of two platforms
  Axes of the two C pairs are parallel, i.e. $C_1 \parallel C_2$

(2) Select an arbitrary point o' on the movable platform as the base point, as shown in Fig. 6.3.

(3) Determine POC set of branches.

According to Eq. (4.4) in Chap. 4, POC set of each branch can be obtained as

$$M_{b_i} = \begin{bmatrix} t^1(\parallel P_i)\ t^1(\perp R_i) \\ r^1(\parallel R_i) \end{bmatrix}, \quad i = 1, 2.$$

(4) Determine number of independent displacement equations of the independent loop

According to topological structure ($C_1 \parallel C_2$) and Eq. (6.8b), there is

$$\begin{aligned} \xi_L &= \dim\{M_{b_1} \cup M_{b_2}\} \\ &= \dim\left\{\begin{bmatrix} t^1(\parallel P_1)\ t^1(\perp R_1) \\ r^1(\parallel R_1) \end{bmatrix} \cup \begin{bmatrix} t^1(\parallel P_2)\ t^1(\perp R_2) \\ r^1(\parallel R_2) \end{bmatrix}\right\} \\ &= \dim\left\{\begin{bmatrix} t^1(\parallel P_1)\ \{t^1(\perp R_1)\}\ t^1(\perp R_2) \\ r^1(\parallel R_1) \end{bmatrix}\right\} = 3 \end{aligned}$$

In the above equation, POC set of R_1 has only one independent element (Table 3.1 in Chap. 3). The non-independent element is put into a brace.

(5) Calculate DOF of the mechanism

According to Eq. (6.8a), DOF of the parallel mechanism is

$$F = \sum_{i=1}^{m} f_i - \sum_{j=1}^{v} \xi_{L_j} = 4 - 3 = 1$$

(6) Determine inactive pairs

In order to determine whether R_1 (take R_1 as an example) is an inactive pair, lock it and a new mechanism is obtained whose first branch is $SOC\{-P_1-\}$. According to Eq. (4.4) in Chap. 4, POC set of the first branch is

$$M_{b_1} = \begin{bmatrix} t^1(\| P_1) \\ r^0 \end{bmatrix}$$

Now, let us determine DOF of the new mechanism.

(a) Determine number of independent displacement equations of the new mechanism
According to topological structure $(P_1 \| C_2)$ and Eq. (6.8b), there is

$$\begin{aligned} \xi_L &= \dim\{M_{b_1} \cup M_{b_2}\} \\ &= \dim\left\{\begin{bmatrix} t^1(\| P_1) \\ r^0 \end{bmatrix} \cup \begin{bmatrix} t^1(\| P_2)\ t^1(\perp R_2) \\ r^1(\| R_2) \end{bmatrix}\right\} \\ &= \dim\left\{\begin{bmatrix} t^1(\| P_1)\ \{t^1(\perp R_2)\} \\ r^1(\| R_2) \end{bmatrix}\right\} = 2 \end{aligned}$$

In this equation, POC set of R_2 has only one independent element (Table 3.1 in Chap. 3). The non-independent element is put into a brace.

(b) Calculate DOF of the new mechanism
According to Eq. (6.8a), DOF of the new mechanism is

$$F^* = \sum_{i=1}^{m} f_i - \sum_{j=1}^{v} \xi_{L_j} = 3 - 2 = 1$$

Since DOF of the new mechanism is equal to DOF of the original mechanism ($F^* = F$), according to the criterion for determination of inactive pairs in the Sect. 6.3.1, R_1 is an inactive pair. Similarly, it is easy to prove that R_2 is also an inactive pair.

(7) Determine POC set of the parallel mechanism

According to Eq. (5.3) in Chap. 5, POC set of the parallel mechanism without inactive pairs shown in Fig. 6.3 is

$$M_{Pa} = M_{b_1} \cap M_{b_2} = \begin{bmatrix} t^1(\| P_1) \\ r^0 \end{bmatrix} \cap \begin{bmatrix} t^1(\| P_2) \\ r^0 \end{bmatrix} = \begin{bmatrix} t^1(\| P_1) \\ r^0 \end{bmatrix}$$

So, movable platform of the parallel mechanism has only one translation.

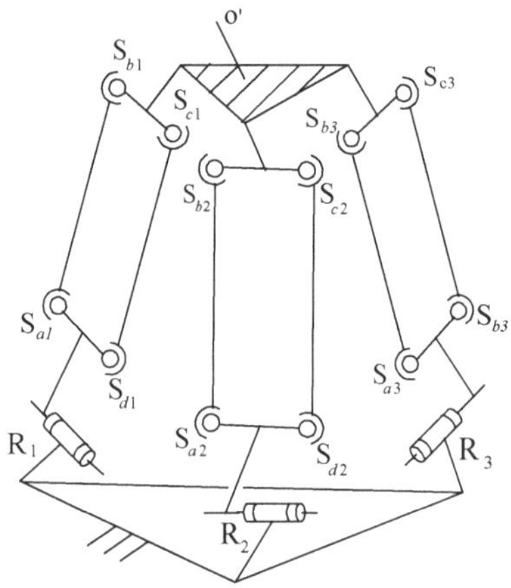

Fig. 6.4 $3 - HSOC\{-R - \diamondsuit(S,S,S,S)-\}$Delta PM

It should be noted that the DOF calculation may be simplified if the base point o′ lies on axis of C_2.

From this example, we know that calculation of PM's POC set always involves calculation of DOF and determination of inactive pairs.

Example 6.3 Determine DOF of the Delta mechanism in Fig. 6.4.

(1) Topological structure

- Topological structure of branches

  Three identical HSOC branches:

  $HSOC\{-R_i - \diamondsuit(S_{a_i}, S_{b_i}, S_{c_i}, S_{d_i})-\}, i = 1,2,3,$ as shown in Fig. 6.4.
  Axis of R_i is parallel to center line of S_{a_i} and S_{d_i} (expressed as $R_i \parallel (a_i d_i)$).
  $\diamondsuit(S_{a_i}, S_{b_i}, S_{c_i}, S_{d_i})$ means that center points of $S_{a_i}, S_{b_i}, S_{c_i}$ and S_{d_i} form a parallelogram upon assembling

- Topological structure of two platforms

  Fixed platform: Axes of R_{11}, R_{21} and R_{31} form a triangle (Fig. 6.4).

  Movable platform: three center lines $(a_i d_i)$(i = 1, 2, 3) form a triangle (Fig. 6.4).

(2) Select an arbitrary point o' on the movable platform as the base point.

(3) Determine POC set of the HSOC branch.

Substituting the POC set of $\diamondsuit(S_{a_i}, S_{b_i}, S_{c_i}, S_{d_i})$ in Table 5.1 in Chap. 5 and the POC set (Eq. (3.2) in Chap. 3) of R pair into Eq. (4.4) in Chap. 4 and considering $R_i \parallel (a_i d_i)$, POC set of the HSOC branch can be obtained as

$$M_{b_i} = \begin{bmatrix} t^1(\perp R_i) \\ r^1(\| R_i) \end{bmatrix} \cup \begin{bmatrix} t^1(\|^{\perp} (a_i d_i)) \cup t^1(\perp(a_i d_i)) \\ r^1(\| (a_i d_i)) \cup r^1(\| (b_i d_i)) \end{bmatrix}$$
$$= \begin{bmatrix} t^3 \\ r^2(\| \diamondsuit(R_i, (b_i d_i))) \end{bmatrix}$$

Here, considering the rotation around the diagonal $(b_i d_i)$, i.e. $r^1(\| (b_i d_i))$.

(4) Determine number of independent displacement equations of each independent loop

(a) According to topological structure $(R_1 \nparallel R_2)$ of the first independent loop and Eq. (6.8b), there is

$$\xi_{L_1} = \dim\{M_{b_1} \cup M_{b_2}\} = \dim\left\{ \begin{bmatrix} t^3 \\ r^2(\| \diamondsuit(R_1, (b_1 d_1)) \end{bmatrix} \cup \begin{bmatrix} t^3 \\ r^2(\| \diamondsuit(R_2, (b_2 d_2)) \end{bmatrix} \right\}$$
$$= \dim.\left\{ \begin{bmatrix} t^3 \\ r^3 \end{bmatrix} \right\} = 6$$

(b) According to the topological structure (R_{11}, R_{21} and R_{31} are not parallel) and Eq. (6.8b), there is

$$\xi_{L_2} = \dim\{(M_{b_1} \cap M_{b_2}) \cup M_{b_3}\}$$
$$= \dim\left\{ \left(\begin{bmatrix} t^3 \\ r^2(\| \diamondsuit(R_1, (b_1 d_1)) \end{bmatrix} \cap \begin{bmatrix} t^3 \\ r^2(\| \diamondsuit(R_2, (b_2 d_2)) \end{bmatrix} \right) \cup \begin{bmatrix} t^3 \\ r^2(\| \diamondsuit(R_3, (b_3 d_3))) \end{bmatrix} \right\}$$
$$= \dim\left\{ \begin{bmatrix} t^3 \\ r^1(\| (\diamondsuit(R_1, (b_1 d_1)) \cap \diamondsuit(R_2, (b_2 d_2)))) \end{bmatrix} \cup \begin{bmatrix} t^3 \\ r^2(\| \diamondsuit(R_3, (b_3 d_3))) \end{bmatrix} \right.$$
$$= \dim\left\{ \begin{bmatrix} t^3 \\ r^3 \end{bmatrix} \right\} = 6$$

(5) Calculate DOF of the mechanism

According to Eq. (6.8a), DOF of the parallel mechanism is

$$F = \sum_{i=1}^{m} f_i - \sum_{j=1}^{2} \xi_{L_j} = 15 - (6 + 6) = 3$$

(6) Determine inactive pairs

In order to determine whether the rotation around the diagonal $(b_3 d_3)$ (take the diagonal $(b_3 d_3)$ as an example) is an inactive pair, lock this rotation and a new mechanism is obtained whose third branch does not have this rotation any longer.

According to Eq. (4.4) in Chap. 4, POC set of the third branch is

$$M_{b_3} = \begin{bmatrix} t^3 \\ r^1(\| R_3) \end{bmatrix}$$

Now, let us determine DOF of the new mechanism.

(a) Determine number of independent displacement equations of the new mechanism.
According to the topological structure and Eq. (6.8b), there are

$$\xi_{L_1} = 6$$

$$\begin{aligned}
\xi_{L_2} &= \dim\{(M_{b_1} \cap M_{b_2}) \cup M_{b_3}\} \\
&= \dim\left\{\left(\begin{bmatrix} t^3 \\ r^2(\| \diamond(R_1, (b_1d_1)) \end{bmatrix} \cap \begin{bmatrix} t^3 \\ r^2(\| \diamond(R_2, (b_2d_2)) \end{bmatrix}\right) \cup \begin{bmatrix} t^3 \\ r^1(\| R_3) \end{bmatrix}\right\} \\
&= \dim\left\{\begin{bmatrix} t^3 \\ r^1(\| (\diamond(R_1, (b_1d_1)) \cap \diamond(R_2, (b_2d_2)))) \end{bmatrix} \cup \begin{bmatrix} t^3 \\ r^1(\| R_3) \end{bmatrix}\right\} \\
&= \dim\left\{\begin{bmatrix} t^3 \\ r^2(\| \diamond\{[\diamond(R_1, (b_1d_1)) \cap \diamond(R_2, (b_2d_2))], R_3\}) \end{bmatrix}\right\} = 5
\end{aligned}$$

(b) Calculate DOF of the new mechanism
According to Eq. (6.8a), DOF of the new mechanism is

$$F^* = \sum_{i=1}^{m} f_i - \sum_{j=1}^{2} \xi_{L_j} = 14 - (6 + 5) = 3$$

Since DOF of the new mechanism is equal to DOF of the original mechanism ($F^* = F$), according to the criterion for determination of inactive pairs in the Sect. 6.3.1, the rotation around the diagonal (b_3d_3) is an inactive DOF (i.e. there is no such rotation). Similarly, it is easy to prove that there are no such rotations around diagonals (b_1d_1) and (b_2d_2).

(7) Determine POC set of the parallel mechanism

According to Eq. (5.3) in Chap. 5, POC set of this parallel mechanism without inactive pairs is

$$\begin{aligned}
M_{Pa} &= M_{b_1} \cap M_{b_2} \cap M_{b_3} \\
&= \begin{bmatrix} t^3 \\ r^1(\| R_1) \end{bmatrix} \cap \begin{bmatrix} t^3 \\ r^1(\| R_2) \end{bmatrix} \cap \begin{bmatrix} t^3 \\ r^1(\| R_3) \end{bmatrix} = \begin{bmatrix} t^3 \\ r^0 \end{bmatrix}
\end{aligned}$$

(8) Selection of driving pairs

According to criteria for selection of driving pairs (Sect. 6.3.2), it is easy to determine that the three R pairs on the fixed platform can be selected as driving pairs simultaneously.

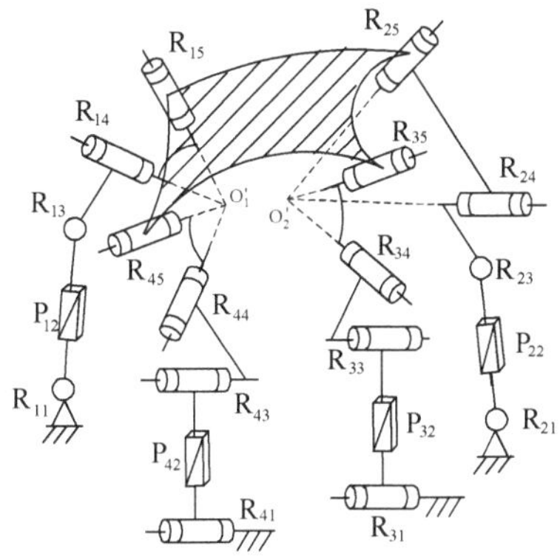

Fig. 6.5 A four-branch parallel mechanism

Example 6.1 Determine DOF of the parallel mechanism in Fig. 6.5 and POC set of its movable platform

(1) Topological structure

- Topological structure of branches:

$$SOC\{-R_{i1}(\perp P_{i2}) \parallel R_{i3} - \overbrace{R_{i4}R_{i5}} -\}, \quad \text{i} = 1,2,3,4.$$

- Topological structure of two platforms

  Fixed platform: $R_{11} \parallel R_{21}$, $R_{31} \parallel R_{41}$, R_{11} and R_{31} are skew in space.

  Movable platform: Axes of R_{14}, R_{15}, R_{44} and R_{45} intersect at point o_1'. Axes of R_{24}, R_{25}, R_{34} and R_{35} intersect at point o_2'.

(2) Select point o_1' on the movable platform as the base point (origin of the moving coordinate system), as shown in Fig. 6.5.

(3) Determine POC set of branches.

   (a) Determine POC set of the first (or the fourth) branch
   According to Example 4.5 in Chap. 4, POC set of the first or the fourth branch is

$$M_{b_i} = \begin{bmatrix} t^2(\perp R_{i1}) \\ r^3 \end{bmatrix}, i = 1,4.$$

   (b) Determine POC set of the second (or the third) branch
   Substitute POC set (Eq. 3.2) of R pair and POC set (Eq. 3.1) of P pair into Eq. (4.4) in Chap. 4, POC set of the second or the fourth branch (relative to base point o_1') is

$$M_{b_i} = \begin{bmatrix} t^3 \\ r^3 \end{bmatrix}, \quad i = 2,3.$$

(4) Determine number of independent displacement equations of each independent loop

(a) According to topological structure of the first independent loop $(R_{11} \nparallel R_{41})$ and Eq. (6.8b), there is

$$\begin{aligned}\xi_{L_1} &= \dim\{M_{b_1} \cup M_{b_4}\} = \dim\left\{\begin{bmatrix} t^2(\perp R_{11}) \\ r^3 \end{bmatrix} \cup \begin{bmatrix} t^2(\perp R_{41}) \\ r^3 \end{bmatrix}\right\} \\ &= \dim\left\{\begin{bmatrix} t^3 \\ r^3 \end{bmatrix}\right\} = 6\end{aligned}$$

(b) According to Eq. (6.8b), there is

$$\begin{aligned}\xi_{L_2} &= \dim\{(M_{b_1} \cap M_{b_4}) \cup M_{b_3}\} \\ &= \dim\left\{\left(\begin{bmatrix} t^2(\perp R_{11}) \\ r^3 \end{bmatrix} \cap \begin{bmatrix} t^2(\perp R_{41}) \\ r^3 \end{bmatrix}\right) \cup \begin{bmatrix} t^3 \\ r^3 \end{bmatrix}\right\} \\ &= \dim\left\{\begin{bmatrix} t^1(\perp(R_{11}, R_{41})) \\ r^3 \end{bmatrix} \cup \begin{bmatrix} t^3 \\ r^3 \end{bmatrix}\right\} = \dim\left\{\begin{bmatrix} t^3 \\ r^3 \end{bmatrix}\right\} = 6\end{aligned}$$

(c) According to Eq. (6.8b), there is

$$\begin{aligned}\xi_{L_3} &= \dim\{(M_{b_1} \cap M_{b_4} \cap M_{b_3}) \cup M_{b_2}\} \\ &= \dim\left\{\left(\begin{bmatrix} t^2(\perp R_{11}) \\ r^3 \end{bmatrix} \cap \begin{bmatrix} t^2(\perp R_{41}) \\ r^3 \end{bmatrix} \cap \begin{bmatrix} t^3 \\ r^3 \end{bmatrix}\right) \cup \begin{bmatrix} t^3 \\ r^3 \end{bmatrix}\right\} \\ &= \dim\left\{\begin{bmatrix} t^3 \\ r^3 \end{bmatrix}\right\} = 6\end{aligned}$$

(5) Calculate DOF of the mechanism

According to Eq. (6.8a), DOF of the parallel mechanism is

$$F = \sum_{i=1}^{m} f_i - \sum_{j=1}^{3} \xi_{Lj} = 20 - (6+6+6) = 2$$

(6) Determine inactive pairs

According to criteria for determination of inactive pairs (Sect. 6.3.1), it is not difficult to determine that this mechanism does not contain any inactive pair.

(7) Determine POC set of the parallel mechanism

According to Eqs. (5.3) and (5.3c) in Chap. 5, POC set of the parallel mechanism is

$$
\begin{aligned}
M_{Pa} &= \begin{bmatrix} t^2(\perp R_{11}) \\ r^3 \end{bmatrix} \cap \begin{bmatrix} t^3 \\ r^3 \end{bmatrix} \cap \begin{bmatrix} t^3 \\ r^3 \end{bmatrix} \cap \begin{bmatrix} t^2(\perp R_{41}) \\ r^3 \end{bmatrix} \\
&= \begin{bmatrix} t^1(\perp(R_{11}, R_{41})) \\ r^1(o_1 - o_2)\ r^2 \end{bmatrix}
\end{aligned}
$$

According to this POC set, moving coordinate system on the movable platform (point o_1' is the origin) has three rotations. In addition, there is also one translation along the common perpendicular of axes of R_{11} and R_{41}. But DOF of the mechanism is two. So, only two of the four elements in the POC set are independent elements. The other two elements are non-independent elements (put into braces).

Obviously, there are several schemes for selection of non-independent elements. One possible scheme is

$$
M_{Pa} = \begin{bmatrix} t^1(\perp(R_{11}, R_{41})) \\ r^1(o_1 - o_2)\ \{r^2\} \end{bmatrix}
$$

The POC set means that the movable platform has one independent rotation, two non-independent rotations (put into braces) and one independent translation along the common perpendicular of axes of R_{11} and R_{41}.

It should be noted that Eq. (6.9) can not be used directly calculate DOF of the parallel mechanism in Fig. 6.5 since the POC set contains non-independent elements.

(8) Select driving pairs

According to criteria for selection of driving pairs (Sect. 6.3.2), it is easy to determine that the four R pairs on the fixed platform can be selected as driving pairs simultaneously.

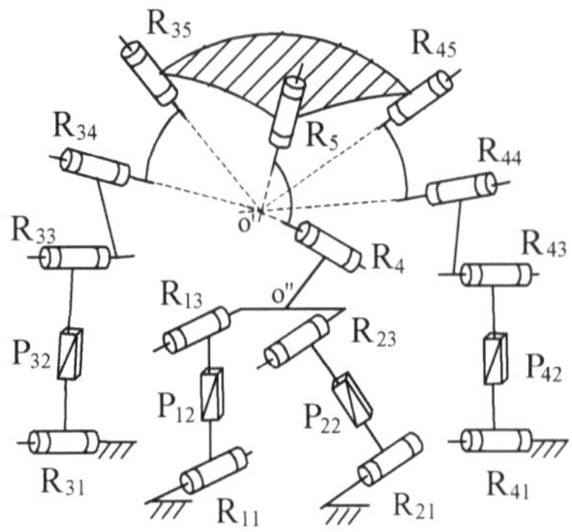

Fig. 6.6 A 3-loop spatial mechanism

Example 6.5 Determine DOF of the 3-loop spatial mechanism in Fig. 6.6.

(1) Topological structure

The 3-loop spatial mechanism in Fig. 6.6 could be considered to be a PM formed by one branch with loop and two simple branches.

- Topological structure of branches:

  The first branch is composed of $SOC\{-\overbrace{R_4R_5}-\}$ and a sub-PM with two sub-branches $SOC\{-R_{i1}(\perp P_{i2})\parallel R_{i3}-\}$(i = 1, 2) as shown in Fig. 6.6. Topological structure of other two branches is

  $SOC\{-R_{j1}(\perp P_{j2})\parallel R_{j3}-\overbrace{R_{j4}R_{j5}}-\}, j=3,4.$

- Topological structure of two platforms:

  Fixed platform: $R_{11}\parallel R_{21}\nparallel R_{31}\parallel R_{41}$.

  Movable platform: $R_4, R_5, R_{34}, R_{35}, R_{44}$ and R_{45} intersect at one common point o′.

(2) Select the base point o′ of the moving platform, as shown in Fig. 6.6.

(3) Determine the POC set of the HSOC branch.

(a) Select the base point o″ of the movable platform of the sub-PM as shown in Fig. 6.6.

(b) Determine POC set of the sub-PM.
According to Eq. (4.4) in Chap. 4, POC set of $SOC\{-R_{i1}(\perp P_{i2})\parallel R_{i3}-\}$ is

$$M_{S_i}=\begin{bmatrix} t^2(\perp R_{i1}) \\ r^1(\parallel R_{i1}) \end{bmatrix},\quad i=1,2$$

(c) Determine the ξ_{L_1} of the first independent loop in sub-PM
According to topological structure $(R_{11}\parallel R_{21})$ and Eq. (6.8b), there is

$$\xi_{L_1} = \dim.\{M_{S_1} \cup M_{S_2}\} = \dim.\left\{\begin{bmatrix} t^2(\perp R_{11}) \\ r^1(\| R_{11}) \end{bmatrix} \cup \begin{bmatrix} t^2(\perp R_{21}) \\ r^1(\| R_{21}) \end{bmatrix}\right\}$$
$$= \dim.\left\{\begin{bmatrix} t^2(\perp R_{11}) \\ r^1(\| R_{11}) \end{bmatrix}\right\} = 3$$

(d) Determine POC set of the sub-PM
According to topological structure $(R_{11} \| R_{21})$ and Eq. (5.3) in Chap. 5, POC set of the sub-PM is

$$M_{Sub-PM} = M_{S_1} \cap M_{S_2} = \begin{bmatrix} t^2(\perp R_{11}) \\ r^1(\| R_{11}) \end{bmatrix} \cap \begin{bmatrix} t^2(\perp R_{21}) \\ r^1(\| R_{21}) \end{bmatrix}$$
$$= \begin{bmatrix} t^2(\perp R_{11}) \\ r^1(\| R_{11}) \end{bmatrix}$$

Since this sub-PM has the same POC set as sub-$SOC\{-R_{11}(\perp P_{12}) \| R_{13}-\}$, it is topologically equivalent to sub-$SOC\{-R_{11}(\perp P_{12}) \| R_{13}-\}$.

(e) Determine the POC set of the HSOC branch.
The first branch is composed of the sub-PM (equivalent to sub-$SOC\{-R_{11}(\perp P_{12}) \| R_{13}-\}$) and $SOC\{-\overset{\frown}{R_4R_5}-\}$ connected in series. According to Eq. (4.4) in Chap. 4, POC set of the HSOC branch is

$$M_{b_1} = \begin{bmatrix} t^2(\perp R_{11}) \\ r^1(\| R_{11}) \end{bmatrix} \cup \begin{bmatrix} t^0 \\ r^2 \| \lozenge(R_4, R_5) \end{bmatrix} = \begin{bmatrix} t^2(\perp R_{11}) \\ r^3 \end{bmatrix}$$

(4) Determine POC sets of other two branches.

According to Example 4.5 in Chap. 4, POC set of these two branches is

$$M_{b_i} = \begin{bmatrix} t^2(\perp R_{i1}) \\ r^3 \end{bmatrix}, \quad i = 3, 4.$$

(5) Determine number of independent displacement equations of each independent loop

(a) According to topological structure $(R_{31} \| R_{41})$ and Eq. (6.8b), there is

$$\xi_{L_2} = \dim\{M_{b_3} \cup M_{b_4}\} = \dim\left\{\begin{bmatrix} t^2(\perp R_{31}) \\ r^3 \end{bmatrix} \cup \begin{bmatrix} t^2(\perp R_{41}) \\ r^3 \end{bmatrix}\right\}$$
$$= \dim\left\{\begin{bmatrix} t^2(\perp R_{31}) \\ r^3 \end{bmatrix}\right\} = 5$$

(b) According to topological structure ($R_{31} \parallel R_{41}$ and $R_{11} \nparallel R_{31}$) and Eq. (6.8b), there is

$$\begin{aligned}\xi_{L_3} &= \dim\{(M_{b_3} \cap M_{b_4}) \cup M_{b_1}\} \\ &= \dim\left\{\left(\begin{bmatrix} t^2(\perp R_{31}) \\ r^3 \end{bmatrix} \cap \begin{bmatrix} t^2(\perp R_{41}) \\ r^3 \end{bmatrix}\right) \cup \begin{bmatrix} t^2(\perp R_{11}) \\ r^3 \end{bmatrix}\right\} \\ &= \dim\left\{\begin{bmatrix} t^2(\perp R_{31}) \\ r^3 \end{bmatrix} \cup \begin{bmatrix} t^2(\perp R_{11}) \\ r^3 \end{bmatrix}\right\} = 6\end{aligned}$$

(6) Calculate the DOF of the mechanism.

According to Eq. (6.8a), DOF of the parallel mechanism is

$$F = \sum_{i=1}^{m} f_i - \sum_{j=1}^{v} \xi_{L_j} = 18 - (3 + 5 + 6) = 4$$

(7) Determine inactive joints

According to criteria for determination of inactive pairs (Sect. 6.3.1), it is easy to prove that this mechanism has no inactive pair.

(8) Determine POC set of the parallel mechanism

According to topological structure ($R_{11} \nparallel R_{31}$, $R_{31} \parallel R_{41}$) and Eq. (5.3) in Chap. 5, POC set of the parallel mechanism is

$$\begin{aligned}M_{Pa} &= M_{b_1} \cap M_{b_3} \cap M_{b_4} \\ &= \begin{bmatrix} t^2(\perp R_{11}) \\ r^3 \end{bmatrix} \cap \begin{bmatrix} t^2(\perp R_{31}) \\ r^3 \end{bmatrix} \cap \begin{bmatrix} t^2(\perp R_{41}) \\ r^3 \end{bmatrix} \\ &= \begin{bmatrix} t^1(\perp(R_{11}, R_{31}) \\ r^3 \end{bmatrix}\end{aligned}$$

According to this POC set, movable platform of this mechanism has three rotations and one translation along the common perpendicular of axes of R_{11} and R_{31}.

(9) Selection of driving pairs

According to criteria for selection of driving pairs (Sect. 6.3.2), it is easy to determine that the four P pairs can be selected as driving pairs simultaneously.

From this example, we learn that the DOF formula is applicable to arbitrary multi-loop spatial mechanisms if sub-PMs are replaced by topologically equivalent sub-SOCs.

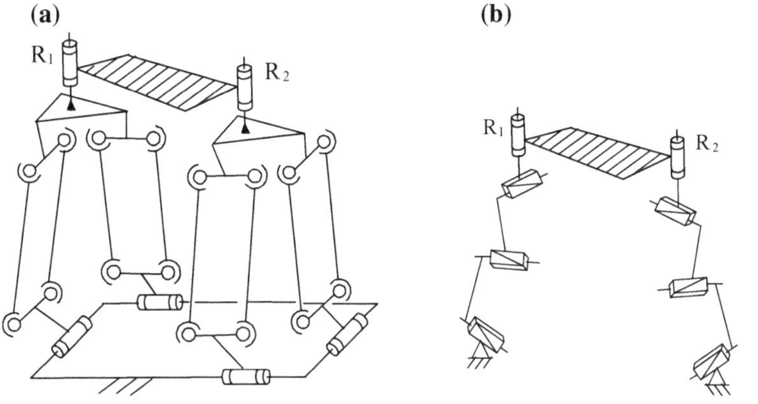

Fig. 6.7 A 7-loop parallel mechanism

Example 6.6 Determine DOF of the 7-loop spatial mechanism in Fig. 6.7a.

(1) Topological structure

- Topological structure of branches

  As shown in Fig. 6.7a, the HSOC branch is composed of an R pair and a parallelogram (formed by four S pairs) connected in series. The axis of R pair is parallel to center line of two adjacent S pairs.

- Topological structure of two platforms:

  Fixed platform: as shown in Fig. 6.7a.

  Movable platform: $R_1 \parallel R_2$.

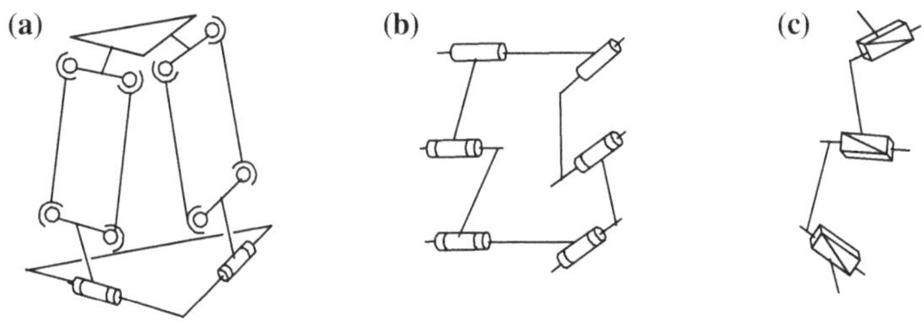

Fig. 6.8 Three topologically equivalent branches

(2) Topologically equivalent branches

According to POC equation for serial mechanisms (Eq. (4.4) in Chap. 4) and POC equation for parallel mechanisms (Eq. (5.3) in Chap. 5), the three branches in Fig. 6.8 have the same POC set $\left(M_b = \begin{bmatrix} t^3 \\ r^0 \end{bmatrix}\right)$. They are topologically equivalent branches. According to topological equivalence principle (Sect. 5.4.3 in Chap. 5), the sub-PM in Fig. 6.7a can be replaced by its equivalent sub-SOC in Fig. 6.8c. Then, the parallel mechanism in Fig. 6.7b is obtained. The two parallel mechanisms in Fig. 6.7 have the same POC set and the same DOF.

Now, let us determine DOF of the parallel mechanism in Fig. 6.7b.

(3) Select an arbitrary point on movable platform as the base point o′.

(4) Determine the POC set of branches

According to Eq. (4.4) in Chap. 4, there is

$$M_{b_j} = \begin{bmatrix} t^3 \\ r^1(\parallel R_j) \end{bmatrix}, \quad j = 1, 2.$$

(5) Determine DOF of the mechanism

According to topological structure $(R_1 \parallel R_2)$ and Eq. (6.8b), there is

$$\begin{aligned} \xi_{L_1} &= \dim\{M_{b_1} \cup M_{b_2}\} = \dim\left\{\begin{bmatrix} t^3 \\ r^1(\parallel R_1) \end{bmatrix} \cup \begin{bmatrix} t^3 \\ r^1(\parallel R_2) \end{bmatrix}\right\} \\ &= \dim\begin{bmatrix} t^3 \\ r^1(\parallel R_1) \end{bmatrix} = 4 \end{aligned}$$

DOF of the mechanism in Fig. 6.7b is

$$F = \sum_{i=1}^{m} f_i - \sum_{j=1}^{v} \xi_{L_j} = 8 - 4 = 4$$

So, DOF of the 7-loop spatial mechanism in Fig. 6.7a is 4.

From this example, we learn that the DOF formula is applicable to arbitrary multi-loop spatial mechanisms if sub-PMs are replaced by topologically equivalent sub-SOCs. DOF calculation can be greatly simplified.

6.5 Summary

(1) The general DOF formula for parallel mechanisms (Eq. 6.8) is proposed based on the topologically equivalent principle. And DOF Calculation is

based on the POC equation of serial mechanisms and POC equation of PMs.

(2) DOF Calculation of the DOF formula is independent of motion position of a PM and the fixed coordinate system. So it can guarantee that full-cycle DOF is obtained.

(3) Based on the DOF formula, criteria for inactive pair determination (Sect. 6.3.1) and criteria for driving pair selection (Sect. 6.3.2) are proposed.

(4) Based on the DOF formula, formulas for determining number of independent displacement equations (Eq. 6.10), number of overconstraints (Eq. 6.11) and redundancy (Eq. 6.12) of the mechanism are derived.

(5) By replacing sub-PMs with topologically equivalent sub-SOCs, DOF of general multi-loop spatial mechanisms can be calculated with this DOF formula very easily (Examples 6.5 and 6.6).

(6) This DOF formula can be used for calculating DOF of PMs whose POC set contains non-independent element (Example 6.4).

(7) POC set calculation of a PM always involves determination of inactive pair and DOF calculation of the mechanisms (Examples 6.2 and 6.3).

Generally, the DOF formula in this chapter reveals the internal relations among topological structure, DOF and POC set of the mechanism and provides a theoretical basis for topology design of parallel mechanisms (Chap. 9).

References

1. Gogu G (2005) Mobility of mechanisms: a critical review. Mech Mach Theory 40:1068–1097
2. Zhao J-S, Zhou K, Feng Z-J (2004) A theory of degrees of freedom for mechanisms. Mech Mach Theory 39:621–643
3. Dai JS, Huang Z, Lipkin H (2004) Screw system analysis of parallel mechanisms and applications to constraint and mobility study. DETC2004-57604, 28th ASME biennial mechanisms and robotics conference, Salt Lake City, USA
4. Dai JS, Huang Z, Lipkin H (2006) Mobility of overconstrained parallel mechanisms. ASME J Mech Des 128:220–229
5. Kong X, Gosselin CM (2005) Mobility analysis of parallel mechanisms based on screw theory and the concept of equivalent serial kinematic chain. In: Proceedings of the 2005 ASME design engineering technical conference and computers and information in engineering conference, DETC2005-85337
6. Kong X, Gosselin CM (2007) Type synthesis of parallel mechanisms. Springer tracts in Advanced Robotics, vol 33
7. Huang Z, Yongsheng Z, Tieshi Z (2006) Higher spatial mechanism theory. Higher Education Press, Beijing
8. Huang Z, Li Q-C, Ding H-F (2013) Theory of parallel mechanisms. Springer, Berlin

9. Gogu G (2005) Mobility and spatiality of parallel robots revisited via theory of linear transformations. Eur J Mech A/Solid 24:690–711
10. Rico JM, Aguilera LD et al (2006) A more general mobility criterion for parallel platforms. Trans ASME J Mech Des 128:207–219
11. Yang T-L, Sun D-J (2006) General formula of degree of freedom for parallel mechanisms and its application. In: Proceedings of ASME 2006 mechanisms conference, DETC2006-99129
12. Yang T-L, Sun D-J (2008) Rank and mobility of single loop kinematic chains. In: Proceedings of the ASME 32-th mechanisms and robots conference. DETC2008-49076
13. Yang T-L, Sun D-J (2008) A general formula of degree of freedom for parallel mechanisms. In: Proceedings of the ASME 32-th mechanisms and robots conference. DETC2008-49077
14. Yang T-L, Sun D-J (2012) A general DOF formula for parallel mechanisms and multi-loop spatial mechanisms. ASME J Mech Robot 4:011001-1–17
15. Yang T-L, Shen H-P, Liu A-X, Dai JS (2015) Review of the formulas for degrees of freedom in the past ten years. Chin J Mech Eng 51(3):69–80
16. Yang T-L (1996) Basic theory of mechanical system: structure, kinematic and dynamic. China Machine Press, Beijing

Chapter 7
Mechanism Composition Principle Based on Single-Open-Chain Unit

7.1 Introduction

As we know, mechanism composition principle and topological structure analysis are the bases for establishment of topology, kinematics and dynamics of mechanisms. During the evolvement of mechanisms theory, three types of mechanism composition principles have been established, i.e. the composition principle based on Assur kinematic chain (AKC) [1–10], the composition principle based on link-pair unit [8, 11–17] and the composition principle based on loop unit [8, 18–25]. Based on each type of composition principle, the corresponding topological structure synthesis methods and modular kinematic and dynamic analysis methods are established.

7.1.1 Mechanism Composition Principle Based on AKC

In 1913, Russian researcher Assur proposed the composition principle for planar mechanisms: any planar mechanism can be regarded as being composed of DOF driving pairs and several planar AKCs [1, 2]. A planar AKC is defined as such a planar kinematic chain (KC) whose DOF is zero and DOF of whose any sub-KC is greater than zero.

An spatial AKC is defined as such a spatial kinematic chain (KC) that its DOF is zero and DOF of its any sub-KC is greater than zero. If one link is removed for an spatial AKC, then an spatial Assur group will be obtained. For the same AKC, a different Assur group will be obtained if a different link is removed.

Therefore, the composition principle for planar mechanism was extended to spatial mechanisms: one spatial mechanism can be regarded as being composed of DOF driving pairs and several spatial AKCs [27], i.e.

T.-L. Yang et al., *Topology Design of Robot Mechanisms*,
https://doi.org/10.1007/978-981-10-5532-4_7

$$KC[F, v] = J_d[F] + \sum AKC_i[v_i] \quad (7.1)$$

where, $KC[F, v]$—a spatial kinematic chain with DOF = F and v independent loops, $J_d[F]$—F driving pairs, $AKC_i[v_i]$—the ith AKC with v_i independent loops.

For example, the 4-branch parallel mechanism with DOF = 4 shown in Fig. 7.1a is composed of 4 driving pairs (Fig. 7.1b) and 2 AKCs (Fig. 7.1c, d), i.e.

$$KC[F = 4, v = 3] = J_d[F = 4] + AKC_1[v_1 = 1] + AKC_2[v_2 = 2]$$

This composition principle has the following features:

(1) There are only finite kinds of v -loop planar AKCs, i.e. one kind for v = 1, one kind for v = 2, 3 kinds for v = 3 and 28 kinds for v = 4 (see Table 7.2) [26]. But all planar mechanisms with practical use contain only 5 kinds of planar AKCs (No. 1–5 in Table 7.2).

(2) Kinematic (dynamic) analysis of a mechanism can be reduced to kinematic (dynamic) analysis of AKCs contained in the mechanism.

(3) AKC is the smallest unit whose kinematic and dynamic analysis can be solved independently.

(4) Based on this composition principle, structure analysis and synthesis method and modular kinematic and dynamic analysis method of mechanisms based on AKC are established [1–10].

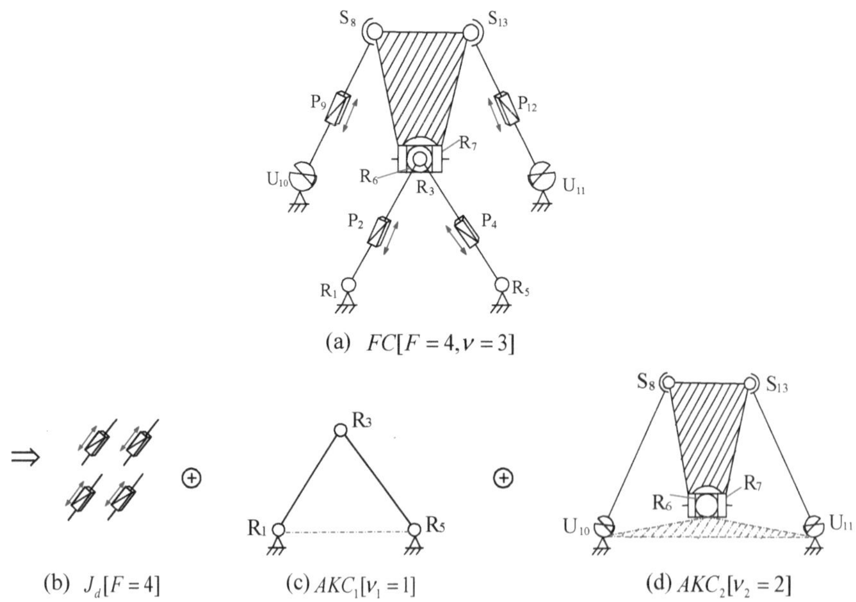

Fig. 7.1 A 4-branch parallel mechanism decomposed into 4 driving pairs and 2 AKCs

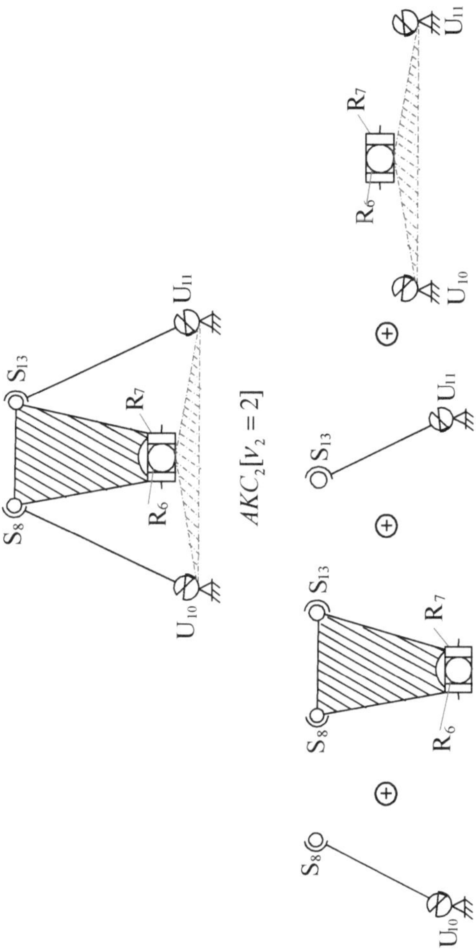

Fig. 7.2 An AKC decomposed into link-pair units

7.1.2 Mechanism Composition Principle Based on Link-Pair Unit

Basic idea of this composition principle is that any kinematic chain can be regarded as being composed of n links connected by m kinematic pairs [8, 11–17], i.e.

$$KC[F, v] = \sum_i link_i + \sum_j Jo\,\mathrm{int}_j \tag{7.2}$$

where, $KC[F, v]$—a kinematic chain with DOF = F and v independent loops.

Obviously, any AKC can also be regarded as being composed of several links connected by kinematic pairs, i.e.

$$AKC[v] = \sum_i link_i + \sum_j Jo\,\mathrm{int}_j$$

For example, the AKC_2 [v = 2] shown in Fig. 7.1d can be regarded as being composed of 4 links connected by 6 kinematic pairs, as shown in Fig. 7.2.

Based on this composition principle, structure analysis and synthesis method and modular kinematic and dynamic analysis method of planar mechanisms based on link-pair unit are established [8, 11–17].

7.1.3 Mechanism Composition Principle Based on Loop Unit

Basic idea of this composition principle is that any kinematic chain with v independent loops can be regarded as being composed of v loops [8, 18–25], i.e.

$$KC[v] = \sum_v loop_i \tag{7.3}$$

where, $KC[v]$—a kinematic chain with v independent loops, $loop_i$ is the ith independent loop.

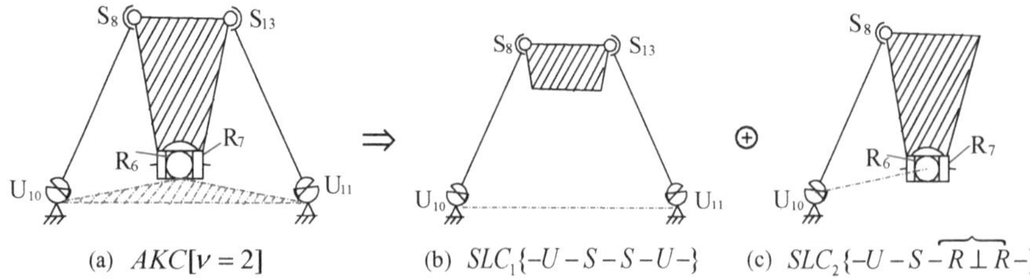

Fig. 7.3 An AKC decomposed into loop units

Obviously, any AKC can also be regarded as being composed of v loop, i.e.

$$AKC[v] = \sum_{v} loop_i$$

For example, the AKC_2 [v = 2] shown in Fig. 7.1d can be regarded as being composed of 2 loops, as shown in Fig. 7.3, i.e.

$$AKC[v=2] = SLC_1\{-U-S-S-U-\} \oplus SLC_2\left\{-U-S-\widehat{R \perp R}-\right\}$$

According to this composition principle, structure analysis and synthesis method and modular kinematic and dynamic analysis method of planar mechanisms based on loop unit are established [8, 18–25].

The following points can be arrived at for the above three composition principles:

(1) Any mechanism is composed of several AKCs (Eq. 7.1). Kinematic (dynamic) analysis of a mechanism can be converted to kinematic (dynamic) analysis of AKC.

(2) AKC is smallest unit whose kinematic and dynamic analysis can be solved independently.

(3) AKC can be regarded as being formed by link-pair unit (Fig. 7.2) or loop unit (Fig. 7.3), on which the systematic theories for mechanism topology, kinematics and dynamics have been established.

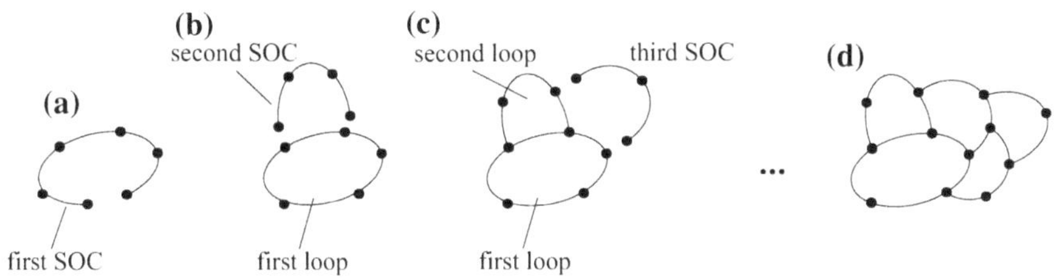

Fig. 7.4 Mechanism composition principle based on SOC unit

7.2 Mechanism Composition Principle Based on SOC Unit

7.2.1 Mechanism Composition Principle

Basic idea of this composition principle: One kinematic chain with v-independent loops can be regarded as being composed of *v* SOCs connected in such a proper order: fix both end links of SOC_1 together to get the first independent loop (Fig. 7.4a, b, where a dot represents a link and an edge represents an kinematic pair), attach two end links of SOC_2 to the first loop to obtain the second independent loop, …, attach two end links of SOC_j to the subchain with (j − 1) independent loops to obtain the jth independent loop, …, attach two end links of SOC_v to the subchain with (v − 1) independent loops to obtain the vth independent loop (Fig. 7.4c, d) [26–32], i.e.

$$KC[F, v] = \sum_{j=1}^{v} SOC_j \tag{7.4}$$

where, $KC[F, v]$—a kinematic chain with DOF = F and v independent loops, SOC_j—the jth SOC.

Obviously, an AKC with v independent loops can also be regarded as being composed of v SOCs, i.e.

$$AKC[v] = \sum_{j=1}^{v} SOC_j \tag{7.5}$$

For example, the AKC_2 shown in Fig. 7.1d can be regarded as being composed of $SOC_1\left\{-U - S - \widehat{R \perp R} -\right\}$ and $SOC_2\{-S - U-\}$, as shown in Fig. 7.5, i.e.

$$AKC_2[v_2 = 2] = SOC_1\left\{-U - S - \widehat{R \perp R} -\right\} + SOC_2\{-S - U-\}$$

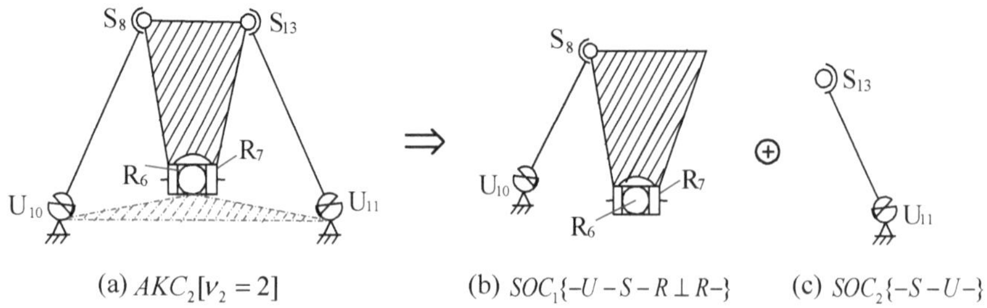

Fig. 7.5 An AKC decomposed into SOC units

Obviously, the parallel mechanism with v independent loops is a simple kind of multi-loop mechanism. It is composed of the $(v + 1)$ SOC branches connected between the fixed platform and the movable platform, as shown in Fig. 7.7: the first two branches and the two platforms form the first independent loop, the third branch connected between the two platforms forms the second independent loop, …, the (j + 1) branch connected between the two platforms forms the jth independent loop, …, the $(v + 1)$ branch connected between the two platforms forms the vth independent loop. This process can be written as follows:

$$PM[F, v] = \sum_{j=1}^{v+1} SOC_j$$

7.2.2 Constraint Degree of SOC

We know from Sect. 7.2.1 that attachment of each SOC branch will add one more independent loop to the kinematic chain. In order to describe the constraint relation of SOC to mechanism DOF, the concept "constraint degree of SOC" is introduced.

(1) Definition of constraint degree of SOC

For a multi-loop spatial mechanism, constraint degree of its SOC_j is defined as [30–33]

$$\Delta_j = \sum_{i=1}^{m_j} f_i - I_j - \xi_{L_j} = \begin{cases} \Delta_j^- = -5, -4, -3, -2, -1. \\ \Delta_j^0 = 0 \\ \Delta_j^+ = +1, +2, +3, \ldots \end{cases} \quad (7.6)$$

where, m_j—the number of kinematic pairs in SOC_j, f_i—DOF of the ith kinematic pair (excluding local DOF), I_j—the number of driving pairs in SOC_j, ξ_{L_j}—the number of independent displacement equations of the jth independent loop which can be obtained by Eq. (6.8b) in Chap. 6 [33].

And there is

$$\sum_{j=1}^{v} \Delta_j = 0 \quad (7.7)$$

So, there is $\sum_{j=1}^{v} \Delta_j^+ = \sum_{j=1}^{v} |\Delta_j^-|$

Equation (7.7) can be proved as follows:

According to Eq. (7.6) and Eq. (6.8a) in Chap. 6, there is

$$\begin{aligned}\sum_{j=1}^{v}\Delta_j &= \sum_{j=1}^{v}\left\{\sum_{i=1}^{m_j} f_i - I_j - \xi_{L_j}\right\} \\ &= \sum_{i=1}^{m} f_i - \sum_{j=1}^{v} I_j - \sum_{j=1}^{v}\xi_{L_j} \\ &= \sum_{i=1}^{m} f_i - F - \sum_{j=1}^{v}\xi_{L_j} \\ &= \left(\sum_{i=1}^{m} f_i - \sum_{j=1}^{v}\xi_{L_j}\right) - F \\ &= F - F = 0\end{aligned}$$

For planar mechanisms, there is $\xi_{L_j} = 3$ and the constraint degree of planar SOCs is [26, 28]

$$\Delta_j = m_j - I_j - 3 = \begin{cases} \Delta_j^- = -2, -1. \\ \Delta_j^0 = 0. \\ \Delta_j^+ = +1, +2, \cdots. \end{cases} \tag{7.8}$$

where, m_j—number of kinematic pairs in SOC_j.

For planar AKCs, there is DOF = 0 ($I_j = 0$). The structure types and constraint degrees of planar SOCs (containing no driving pair) are listed in Table 7.1.

Table 7.1 Structure types and constraint degrees of planar SOCs [26, 28]

Δ_j^-	$\Delta_j^- = -1$ 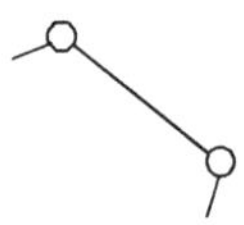	$\Delta_j^- = -2$ 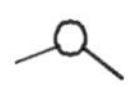
Δ_j^0	$\Delta_j^0 = 0$ 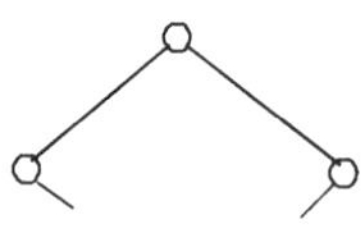	
Δ_j^+	$\Delta_j^+ = +1$ 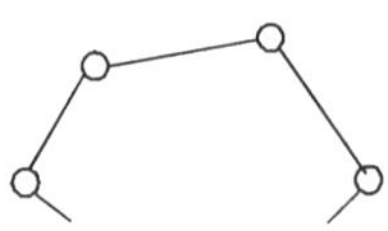	$\Delta_j^+ = +2$ 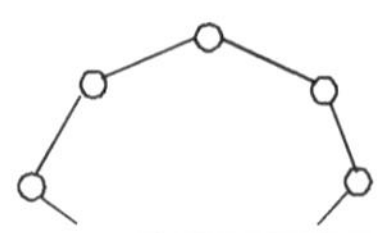

(2) Physical meaning of constraint degree of SOC

Constraint degree of SOC represents constraint relation of SOC to mechanism DOF:

(a) SOC with negative constraint degree ($SOC(\Delta_j^-)$) brings $\left|\Delta_j^-\right|$ constraints to mechanism and reduces DOF of the mechanism by $\left|\Delta_j^-\right|$.
(b) SOC with zero constraint degree ($SOC(\Delta_j^0)$) brings zero constraint to mechanism and does not change DOF of the mechanism.
(c) SOC with positive constraint degree ($SOC(\Delta_j^+)$) increases DOF of the mechanism by Δ_j^+.

7.3 Coupling Degree of AKC

7.3.1 Structure Decomposition of AKC Based on SOC Unit

An AKC with v independent loops can be decomposed into v ordered SOCs in such way: choose the SOC with the smallest constraint degree (i.e. $\Delta_1 = \min\{\sum_{i=1}^{m_1} f_i - I_1 - \xi_{L_1}\}$) among all possible SOCs which can form the first loop of the AKC as SOC_1 and obtain its constraint degree Δ_1, choose the SOC with the smallest constraint degree (i.e. $\Delta_2 = \min\{\sum_{i=1}^{m_2} f_i - I_2 - \xi_{L_2}\}$) among all remaining possible SOCs which can form the second loop of the AKC as SOC_2 and obtain its constraint degree Δ_2, …, choose the SOC with the smallest constraint degree (i.e. $\Delta_j = \min\{\sum_{i=1}^{m_j} f_i - I_j - \xi_{L_j}\}$) among all remaining possible SOCs which can form the jth loop of the AKC as SOC_j and obtain its constraint degree Δ_j, …, obtain SOC_v and its constraint degree Δ_v.

Note: According to this decomposition method, there may be several different decomposition schemes for an AKC.

7.3.2 Coupling Degree of AKC

(1) Definition of coupling degree of AKC

Based on the AKC structure decomposition method described in Sect. 7.3.1, an AKC with v independent loops can be decomposed into v $SOC(\Delta_j)$ $(j = 1, 2, \cdots, v)$ and there may be more than one decomposition schemes.

Based on Eq. (7.7), coupling degree of the AKC is defined as [26–31]

$$\kappa = \frac{1}{2} \min\left\{ \sum_{j=1}^{v} |\Delta_j| \right\} \tag{7.9}$$

where, κ—coupling degree of the AKC, min{•} means that the decomposition scheme with the smallest $\sum |\Delta_j|$ value is selected.

(2) Physical meaning of coupling degree of AKC

(a) Coupling degree represents complexity of the constraints among SOCs in the AKC.

- $\kappa = 0$ means constraints of each SOC are independent and not coupled together.
- $\kappa > 0$ means constraints of each SOC are coupled together. The larger the κ, the stronger the coupling relations among constraints of each SOC and the more complex the AKC's topological structure.

(b) Coupling degree represents complexity of the AKC's kinematic (dynamic) analysis. Kinematic (dynamic) analysis of an AKC is converted to kinematic (dynamic) analysis of ordered SOCs and dimension (number of unknowns) of kinematic analysis equations of these ordered SOCs is equal to κ (refer to Sect. 7.7.2 for detail). So, the larger the κ, the more complex the AKC's kinematic (dynamic) analysis.

Obviously, the coupling degree is the intrinsic property of an AKC. It is a topological invariant.

7.3.3 Examples

Example 7.1 Determine coupling degree of the planar 3-loop AKC shown in Fig. 7.6.

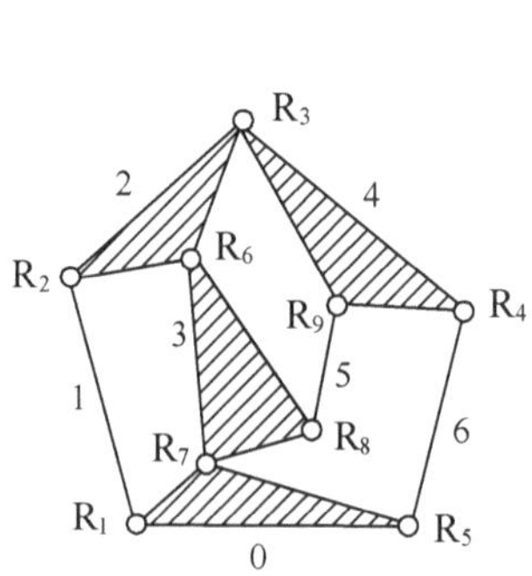

Fig. 7.6 A 3-loop planar AKC

According to the AKC structure decomposition method described in Sect. 7.3.1, the AKC in Fig. 7.6 can be decomposed into three ordered SOCs.

(1) Scheme 1

$$SOC_1\{-R_1R_2R_6R_7-\}, \Delta_1 = m_1 - I_1 - 3 = 4 - 0 - 3 = +1;$$
$$SOC_2\{-R_3R_9R_8-\}, \Delta_2 = 3 - 0 - 3 = 0;$$
$$SOC_3\{-R_4R_5-\}, \Delta_3 = 2 - 0 - 3 = -1.$$
$$k_{(1)} = \frac{1}{2}\{\sum_{j=1}^{v_3} |\Delta_j|\} = 1$$

(2) Scheme 2

$$SOC_1\{-R_6R_3R_9R_8-\}, \Delta_1 = 4 - 0 - 3 = +1;$$
$$SOC_2\{-R_4R_5R_7-\}, \Delta_2 = 3 - 0 - 3 - 0;$$
$$SOC_3\{-R_1R_2-\}, \Delta_3 = 2 - 0 - 3 = -1.$$
$$k_{(2)} = \frac{1}{2}\{\sum_{j=1}^{v_3} |\Delta_j|\} = 1$$

(3) Other decomposition schemes are omitted here, since their coupling degrees are even larger.

(4) Coupling degree of AKC

According to Eq. (7.9), the smaller value of $k_{(1)}$ and $k_{(2)}$ shall be taken as coupling degree of this AKC, i.e.

$$k = \frac{1}{2}\min\{\sum_{j=1}^{v} |\Delta_j|\} = 1$$

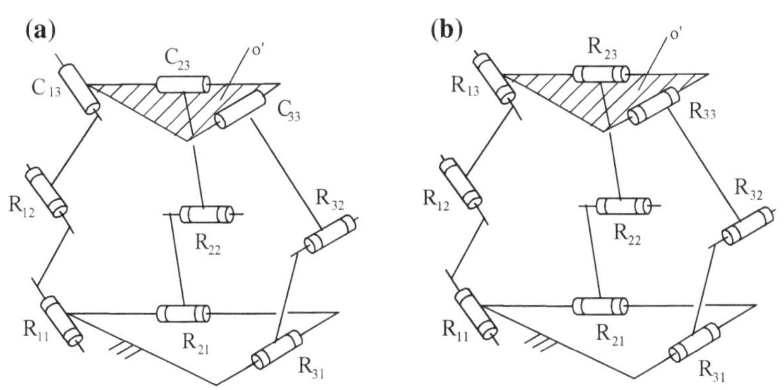

Fig. 7.7 3-$SOC\{-R \parallel R \parallel R-\}$ spatial AKC

Example 7.2 Determine coupling degree of the spatial AKC in Fig. 7.7b contained in the parallel mechanism in Fig. 7.7a (the translation of each C pair is used as the driving input).

(1) Topological structure

(a) Topological structure of branches:

$$SOC\{-R_{i1} \parallel R_{i2} \parallel R_{i3}-\}, i = 1, 2, 3.$$

(b) Topological structure of two platforms
R_{11}, R_{21} and R_{31} are skewed in space, as shown in Fig. 7.7.

(2) Take an arbitrary point on the movable platform as the base point o′.

(3) Determine POC set of the branch
According to Eq. (4.4) in Chap. 4, POC set of each branch is

$$M_{b_i} = \begin{bmatrix} t^2(\perp R_{i1}) \\ r^1(\parallel R_{i1}) \end{bmatrix}, \ i = 1, 2, 3.$$

(4) Determine SOC_1 and Δ_1

① Since each SOC branch has the same topological structure, according to the AKC structure decomposition method described in Sect. 7.3.1, the first SOC can be selected as

$$SOC_1\{-R_{11} \| R_{12} \| R_{13} - R_{23} \| R_{22} \| R_{21}-\}$$

② According to Eq. (6.8b) in Chap. 6, the number of independent displacement equations of the first independent loop is

$$\xi_{L_1} = \dim\{M_{b_1} \cup M_{b_2}\} = \dim\left\{\begin{bmatrix} t^2(\perp R_{11}) \\ r^1(\parallel R_{11}) \end{bmatrix} \cup \begin{bmatrix} t^2(\perp R_{21}) \\ r^1(\parallel R_{21}) \end{bmatrix}\right\}$$
$$= \dim\left\{\begin{bmatrix} t^3 \\ r^2(\parallel \diamondsuit(R_{11}, R_{21})) \end{bmatrix}\right\} = 5$$

③ According to Eq. (7.6), constraint degree of SOC_1 is

$$\Delta_1 = \sum_{i=1}^{(m_1+m_2)} f_i - I_j - \xi_{L_i} = 6 - 0 - 5 = +1$$

(5) Determine SOC_2 and Δ_2

① According to the AKC structure decomposition method described in Sect. 7.3.1, the second SOC can be selected as

$$SOC_2\{-R_{31} \| R_{32} \| R_{33}-\}$$

② According to Eq. (6.8b) in Chap. 6, the number of independent displacement equations of the second independent loop is

$$\begin{aligned}\xi_{L_2} &= \dim\{(M_{b_1} \cap M_{b_2}) \cup M_{b_3}\} \\ &= \dim\left\{\left(\begin{bmatrix} t^2(\perp R_{11}) \\ r^1(\| R_{11}) \end{bmatrix} \cap \begin{bmatrix} t^2(\perp R_{21}) \\ r^1(\| R_{21}) \end{bmatrix}\right) \cup \begin{bmatrix} t^2(\perp R_{31}) \\ r^1(\| R_{31}) \end{bmatrix}\right\} \\ &= \dim\left\{\left(\begin{bmatrix} t^1(\perp(R_{11}, R_{21})) \\ r^0 \end{bmatrix}\right) \cup \begin{bmatrix} t^2(\perp R_{31}) \\ r^1(\| R_{31}) \end{bmatrix}\right\} \\ &= \dim\left\{\begin{bmatrix} t^3 \\ r^1(\| R_{31}) \end{bmatrix}\right\} = 4\end{aligned}$$

③ According to Eq. (7.6), constraint degree of SOC_2 is

$$\Lambda_2 = \sum\nolimits_{i=1}^{m_3} f_i - I_j - \xi_{L_2} = 3 - 0 - 4 = -1.$$

(6) Determine coupling degree of this AKC

According to Eq. (7.9), considering typological symmetry, coupling degree of this AKC is

$$k = \frac{1}{2}\min\left\{\sum_{j=1}^{v=2} |\Delta_j|\right\} = \frac{1}{2}(1 + |-1|) = 1$$

(7) Discussion

The first independent loop is formed by the first SOC branch and the second SOC branch (natural loop) and there is $\xi_{L_2} = 5$. If the second independent loop were formed by second SOC branch and the third SOC branch (natural loop), there would be $\xi_{L_2} = 5$. For these two natural loops, there exists relevance between $\xi_{L_1} = 5$ and $\xi_{L_2} = 5$. So, the second independent loop should be formed by sub-PM (formed by the first two SOC branches) and the third SOC branch (topologically equivalent loop) and there is $\xi_{L_2} = 4$ (see Eq. (6.8b)).

POC set of the sub-PM (formed by the first two SOC branches) is

$$\begin{aligned}M_{pa(1-2)} &= M_{b_1} \cap M_{b_2} \\ &= \begin{bmatrix} t^2(\perp R_{11}) \\ r^1(\| R_{11}) \end{bmatrix} \cap \begin{bmatrix} t^2(\perp R_{21}) \\ r^1(\| R_{21}) \end{bmatrix} = \begin{bmatrix} t^1(\perp(R_{11}, R_{21})) \\ r^0 \end{bmatrix}\end{aligned}$$

This sub-PM is topologically equivalent to $SOC\{-P_{(1-2)}-\}$ branch. The second independent loop is formed by this sub-PM and the third branch (topologically equivalent loop). It is different with the natural loop formed by the second branch and the third branch. So, determination of the number of independent displacement equations of the topologically equivalent loop is a key step in decomposing a mechanism into ordered SOC units.

7.4 The General Method for AKC Determination

Since kinematic (dynamic) analyses of a mechanism can be converted to kinematic (dynamic) analyses of AKCs, the general method for AKC determination is discussed in this section.

7.4.1 *Structure Decomposition of Mechanisms*

A mechanism with v independent loops can be decomposed into v ordered SOCs in such way: choose the SOC with the smallest constraint degree (i.e.$\Delta_1 = \min\{\sum_{i=1}^{m_1} f_i - I_1 - \xi_{L_1}\}$) among all possible SOCs which can form the first loop of the mechanism as SOC_1 and obtain its constraint degree Δ_1, choose the SOC with the smallest constraint degree (i.e. $\Delta_2 = \min\{\sum_{i=1}^{m_2} f_i - I_2 - \xi_{L_2}\}$) among all remaining possible SOCs which can form the second loop of the mechanism as SOC_2 and obtain its constraint degree Δ_2, …, choose the SOC with the smallest constraint degree (i.e. $\Delta_j = \min\{\sum_{i=1}^{m_j} f_i - I_j - \xi_{L_j}\}$) among all remaining possible SOCs which can form the jth loop of the mechanism as SOC_j and obtain its constraint degree Δ_j, …, obtain SOC_v and its constraint degree Δ_v.

Note: According to this decomposition method, there may be several different decomposition schemes for a mechanism.

7.4.2 *Criteria for AKC Determination*

According to the structure decomposition method described in Sect. 7.4.1, a kinematic chain can be decomposed into v ordered SOC_i (i = 1,2, …,v) and constraint degree Δ_i (i = 1,2, …,v) of each SOC_i can be obtained. Divide $\Delta_1, \Delta_2, \ldots, \Delta_v$ into several smallest groups and ensure that the sum of constraint degree of SOCs for each group is zero, as shown in Fig. 7.8. Each such group represents an AKC. Coupling degree of each AKC can be obtained from Eq. (7.9) [26–31].

$$KC[F, v] = \sum_{j=1}^{v} SOC_j(\Delta_j) \Rightarrow \sum_i AKC_i[v_i, \kappa_i] \tag{7.10}$$

Note: There may be several different grouping schemes. The scheme with largest number of groups shall be selected.

$$\Delta_1, \cdots, \Delta_{S_1} \quad \Big| \quad \Delta_{(S_1+1)}, \cdots, \Delta_{S_2} \quad \Big| \quad \cdots\cdots \quad \Big| \quad \Delta_{(S_r+1)}, \cdots, \Delta_v$$

$$\sum_{i=1}^{S_1} \Delta_j = 0 \quad \Big| \quad \sum_{j=(S_1+1)}^{S_2} \Delta_j = 0 \quad \Big| \quad \cdots\cdots \quad \Big| \quad \sum_{j=(S_r+1)}^{v} \Delta_j = 0$$

Fig. 7.8 The smallest group during structure decomposition

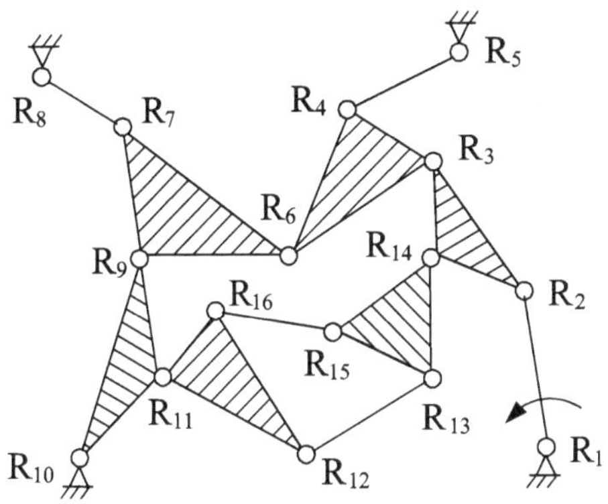

Fig. 7.9 A 5-loop planar mechanism

7.4.3 Examples

Example 7.3 Determine AKCs contained in the planar mechanism in Fig. 7.9 (R_1 is the driving pair) and calculate their coupling degree κ.

According the structure decomposition method described in Sect. 7.4.1 and the AKC determination method in Sect. 7.4.2, there are several different SOC grouping schemes.

(1) Scheme 1

$$\begin{aligned}&SOC_1\{-R_1R_2R_3R_4R_5-\}(\mathrm{R_1 is\ driving\ pair}),\\&\Delta_1 = m_1 - I_1 - 3 = 5 - 1 - 3 = +1;\\&SOC_2\{-R_6R_7R_8-\}, \Delta_2 = 3 - 0 - 3 = 0;\\&SOC_3\{-R_9R_{10}-\}, \Delta_3 = 2 - 0 - 3 = -1.\end{aligned}$$

There is $\sum_{j=1}^{3} \Delta_j = 0$, so the above 3 SOCs correspond to the first AKC:

$$k_{(1)} = \frac{1}{2}\{\sum_{j=1}^{v_3} |\Delta_j|\} = 1$$

$$\begin{aligned}&SOC_4\{-R_{11}R_{12}R_{13}R_{14}-\}, \Delta_4 = 4 - 0 - 3 = +1.\\&SOC_5\{-R_{15}R_{16}-\}, \Delta_5 = 2 - 0 - 3 = -1.\end{aligned}$$

There is $\sum_{j=4}^{5} \Delta_j = 0$, so the above 2 SOCs correspond to the second AKC:

$$k_{(2)} = \frac{1}{2}\{\sum_{j=1}^{v_3} |\Delta_j|\} = 1$$

(2) Scheme 2

$$SOC_1\{-R_8R_7R_9R_{10}-\}, \Delta_1 = m_1 - I_1 - 3 = 4 - 0 - 3 = +1;$$
$$SOC_2\{-R_6R_4R_5-\}, \Delta_2 = 3 - 0 - 3 = 0;$$
$$SOC_3\{-R_1R_2R_3-\}(\mathrm{R_1 is\ driving\ pair}), \Delta_3 = 3 - 1 - 3 = -1.$$

There is $\sum_{j=1}^{3} \Delta_j = 0$, so the above 3 SOCs correspond to the first AKC:

$$k_{(1)} = \frac{1}{2}\{\sum_{j=1}^{v_3} |\Delta_j|\} = 1$$

$$SOC_4\{-R_{11}R_{12}R_{13}R_{14}-\}, \Delta_4 = 4 - 0 - 3 = +1.$$
$$SOC_5\{-R_{15}R_{16}-\}, \Delta_5 = 2 - 0 - 3 = -1.$$

There is $\sum_{j=4}^{5} \Delta_j = 0$, so the above 2 SOCs correspond to the second AKC:

$$k_{(2)} = \frac{1}{2}\{\sum_{j=1}^{v_3} |\Delta_j|\} = 1$$

(3) Scheme 3

$$SOC_1\{-R_{12}R_{13}R_{15}R_{16}-\}, \Delta_1 = m_1 - I_1 - 3 = 4 - 0 - 3 = +1;$$
$$SOC_2\{-R_{11}R_{10}R_1R_2R_{14}-\}(\mathrm{R_1 is\ driving\ pair}), \Delta_2 = 5 - 1 - 3 = +1;$$
$$SOC_3\{-R_3R_6R_9-\}, \Delta_3 = 3 - 0 - 3 = 0.$$
$$SOC_4\{-R_4R_5-\}, \Delta_4 = 2 - 0 - 3 = -1.$$
$$SOC_5\{-R_7R_8-\}, \Delta_5 = 2 - 0 - 3 = -1.$$

There is $\sum_{j=1}^{5} \Delta_j = 0$. Since the number of AKCs is fewer, this scheme is ignored.

(4) Other schemes are omitted here, since the number of AKCs is fewer.

(5) Determine AKCs contained in this mechanism

Both usable schemes indicate that this planar mechanism contains two AKCs. It can be expressed as

$$KC[F = 1, v = 5] = J_d[1] \oplus AKC_1[v = 3, \kappa_1 = 1] \oplus AKC_2[v = 2, \kappa_2 = 1]$$

Example 7.4 Determine AKCs contained in the 3-loop parallel mechanism in Fig. 7.10a (P_2, P_4, P_9, P_{12} are driving pairs) and calculate their coupling degree κ.

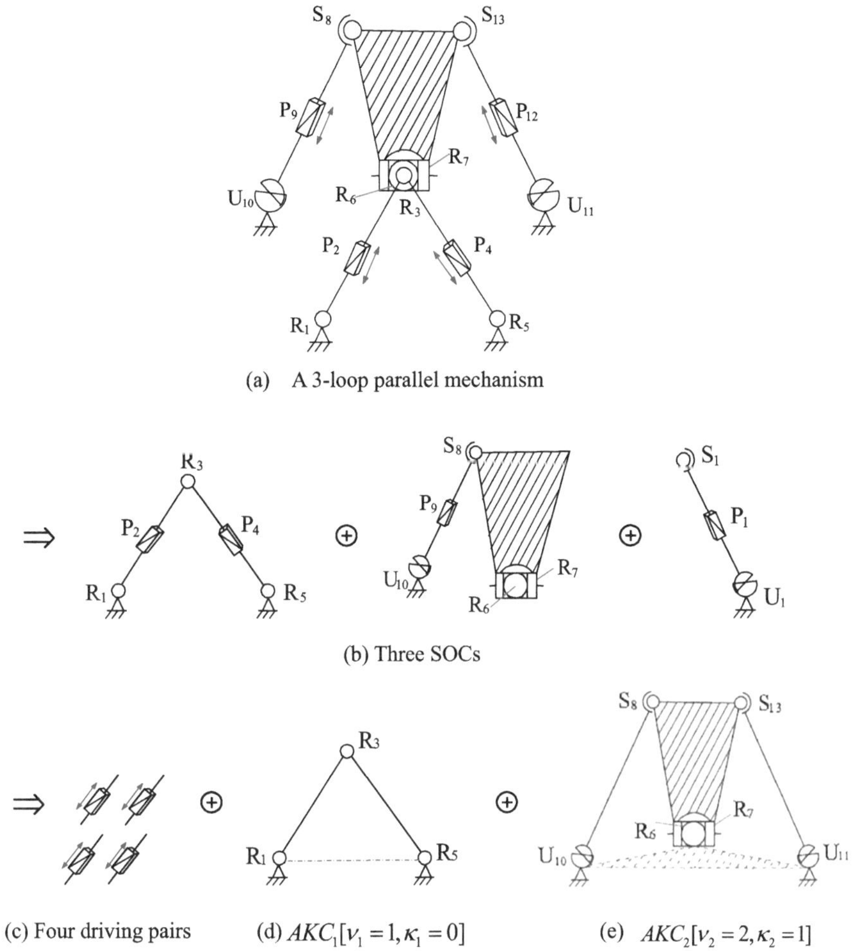

Fig. 7.10 A four DOF 3-loop parallel mechanism

(1) Topological structure

(a) Topological structure of branches:

- One HSOC branch: composed of a sub - $PM\{-R_1(\perp P_2) \parallel R_3(\perp P_4) \parallel R_5-\}$ and a sub - $SOC\{-\overbrace{R_6 \perp R_7}-\}$. There is $R_3|R_6$.
- Two SOC branches: $SOC\{-U_{10}-P_9-S_8-\}$, $SOC\{-U_{11}-P_{12}-S_{13}-\}$.

(b) Topological structure of two platforms

- Movable platform: R_7, S_8 and S_{13} are allocated as shown in Fig. 7.10a.
- Fixed platform: $R_1 \parallel R_5$. U_{10} and U_{11} are allocated as shown in Fig. 7.10a.

(2) Take the intersection of axes of R_6 and R_7 on the movable platform as the base point o′.

(3) Determine POC set of branches

① POC set of the HSOC branch
According to Eq. (5.3) in Chap. 5, POC set of sub - $PM\{-R_1(\perp P_2) \parallel R_3(\perp P_4) \parallel R_5-\}$ is

$$M_{b_{1-1}} = \begin{bmatrix} t^2(\perp R_1) \\ r^0 \end{bmatrix}$$

According to Eq. (4.4) in Chap. 4, POC set of the sub - $SOC\{-\overbrace{R_6 \perp R_7}-\}$ is

$$M_{b_{1-2}} = \begin{bmatrix} t^0 \\ r^2(\parallel \diamondsuit(R_6, R_7)) \end{bmatrix}$$

According to Eq. (4.4), POC set of the HSOC branch is

$$M_{b_1} = M_{1-1} \cup M_{1-2} = \begin{bmatrix} t^2(\perp R_1) \\ r^0 \end{bmatrix} \cup \begin{bmatrix} t^0 \\ r^2(\parallel \diamondsuit(R_6, R_7)) \end{bmatrix}$$
$$= \begin{bmatrix} t^2(\perp R_1) \\ r^2(\parallel \diamondsuit(R_6, R_7)) \end{bmatrix}$$

② POC set of SOC branches
According to Eq. (4.4), POC set of each SOC branch is

$$M_{b_i} = \begin{bmatrix} t^3 \\ r^3 \end{bmatrix}, \mathrm{i} = 2, 3.$$

(4) Determine SOC_1 and Δ_1

① According to the mechanism structure decomposition method described in Sect. 7.4.1 and after comparison, the first SOC should be selected as

$$SOC_1\{-R_1(\perp P_2) \parallel R_3(\perp P_4) \parallel R_5-\}, (\mathrm{P_2 \text{ and } P_4 \text{ are driving pairs}})$$

② Since SOC_1 form a planar five-bar loop, there is $\xi_{L_1} = 3$.

③ According to Eq. (7.8), there is

$$\Delta_1 = \sum_{i=1}^{5} f_i - I_1 - \xi_{L_1} = 5 - 2 - 3 = 0.$$

(5) Determine SOC_2 and Δ_2

① Since the mechanism in Fig. 7.10a has topological symmetry, the second SOC can be selected as being composed of $SOC\{-U_{10} - P_9 - S_8-\}$ and sub - $SOC\{-\overbrace{R_6 \perp P_7}-\}$ connected in series, i.e. $SOC_2\{-U_{10} - P_9 - S_8- \overbrace{R_7 \perp R_6}-\}$, ($P_9$ is a driving pair)

② According to Eq. (6.8b) in Chap. 6, there is

$$\xi_{L_2} = \dim\{M_{b_2} \cup M_{b_{1-2}}\} = \dim\left\{\begin{bmatrix} t^3 \\ r^3 \end{bmatrix} \cup \begin{bmatrix} t^0 \\ r^2(\| \diamondsuit(R_6, R_7)) \end{bmatrix}\right\}$$
$$= \dim\left\{\begin{bmatrix} t^3 \\ r^3 \end{bmatrix}\right\} = 6$$

③ According to Eq. (7.6), there is

$$\Delta_2 = \sum_{i=1}^{5} f_i - I_2 - \xi_{L_2} = 8 - 1 - 6 = +1.$$

(6) Determine SOC_3 and Δ_3

① The third SOC is
$SOC_3\{-U_{11} - P_{12} - S_{13}-\}$, ($P_{12}$ is a driving pair).

② According to Eq. (6.8b) in Chap. 6, there is

$$\xi_{L_3} = \dim\{(M_{b_1} \cap M_{b_2}) \cup M_{b_3}\}$$
$$= \dim\left\{\left(\begin{bmatrix} t^2(\perp R_1) \\ r^2(\| \diamondsuit(R_6, R_7)) \end{bmatrix} \cap \begin{bmatrix} t^3 \\ r^3 \end{bmatrix}\right) \cup \begin{bmatrix} t^3 \\ r^3 \end{bmatrix}\right\}$$
$$= \dim\left\{\begin{bmatrix} t^3 \\ r^3 \end{bmatrix}\right\} = 6$$

③ According to Eq. (7.6), there is

$$\Delta_3 = \sum_{i=1}^{5} f_i - I_3 - \xi_{L_3} = 6 - 1 - 6 = -1$$

So, the parallel mechanism in Fig. 7.10a is composed of three SOCs, as shown in Fig. 7.10b.

(7) Determine AKCs contained in this mechanism

For the three SOCs and their constraint degree Δ_1, Δ_2 and Δ_3, it is easy to determine that this mechanism has two AKCs according to criteria for AKC determination in Sect. 7.4.2.

① The first AKC

$$SOC_1\{-R_1(\perp P_2) \parallel R_3(\perp P_4) \parallel R_5-\}, \Delta_1 = 0. \Rightarrow \sum_{j=1}^{1} \Delta_j = 0,$$

$$\kappa_1 = \tfrac{1}{2}\min\{\sum_{j=1}^{1} |\Delta_j|\} = 0.$$

This SOC corresponds to $AKC_1[v_1 = 1, \kappa_1 = 0]$($P_2$ and P_4 are driving pairs).

② The second AKC.

$$\left.\begin{array}{ll} SOC_2\{-R_6 \perp R_7 - U_8 - P_9 - S_{10}-\}, & \Delta_2 = +1 \\ SOC_3\{-S_{11} - P_{12} - U_{13}-\}, & \Delta_3 = -1 \end{array}\right\} \Rightarrow \sum_{j=2}^{3} \Delta_j = 0,$$

$$\kappa_2 = \frac{1}{2}\min\{\sum_{j=2}^{3} |\Delta_j|\} = 1.$$

These two SOCs correspond to $AKC_2[v_2 = 2, \kappa_2 = 1]$, ($P_9$ and P_{12} are driving pairs).

So, the parallel mechanism in Fig. 7.10a contains two AKCs (Fig. 7.10d, e):

$$KC[F = 4, v = 3] = J_d[4] \oplus AKC_1[v_1 = 1, \kappa_1 = 0] \oplus AKC_2[v_2 = 2, \kappa_2 = 1].$$

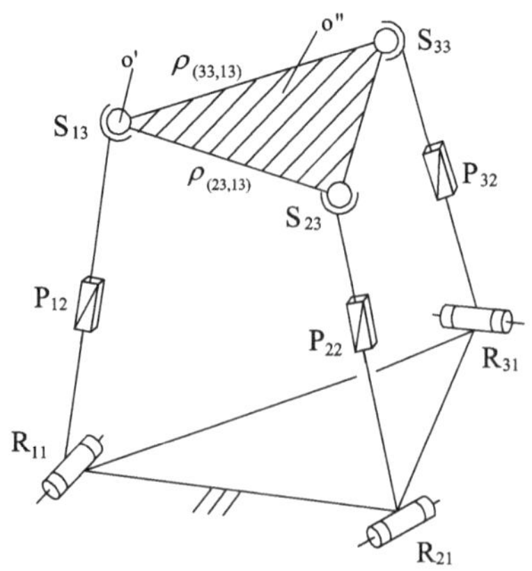

Fig. 7.11 A $3 - SOC\{-R \perp P - S-\}$ PM

Example 7.5 Determine AKCs contained in the parallel mechanism in Fig. 7.11 (P_{12}, P_{22} and P_{32} are driving pairs) and calculate their coupling degree κ.

(1) Topological structure of the mechanism

(a) Topological structure of branches:

$$SOC\{-R_{i1}\perp P_{i2} - S_{i3}-\}, i = 1, 2, 3.$$

(b) Topological structure of two platforms: three R pairs are allocated on the fixed platform as shown in Fig. 7.11.

(2) Take an arbitrary point on the moving platform as the base point o′.

(3) Determine POC set of the branches
According to Eq. (4.4) in Chap. 4, POC set of the SOC branch is

$$M_{b_j} = \begin{bmatrix} t^3 \\ r^3 \end{bmatrix}, j = 1, 2, 3.$$

(4) Determine SOC_1 and Δ_1

① According to the mechanism structure decomposition method described in Sect. 7.4.1, the first SOC is selected as
$SOC_1\{-R_{11}\perp P_{12} - S_{13} - S_{23} - P_{22}\perp R_{21}-\}$, ($R_{11}$ and R_{21} are driving pairs)

② According to Eq. (6.8b) in Chap. 6, there is

$$\begin{aligned} \xi_{L_2} &= \dim\{(M_{b_1} \cap M_{b_2}) \cup M_{b_3}\} \\ &= \dim\left\{\left(\begin{bmatrix} t^3 \\ r^3 \end{bmatrix} \cap \begin{bmatrix} t^3 \\ r^3 \end{bmatrix}\right) \cup \begin{bmatrix} t^3 \\ r^3 \end{bmatrix}\right\} = \dim\left\{\begin{bmatrix} t^3 \\ r^3 \end{bmatrix}\right\} = 6 \end{aligned}$$

③ According to Eq. (7.6), there is

$$\Delta_1 = \sum_{i=1}^{m_1} f_i - I_1 - \xi_{L_1} = 9 - 2 - 6 = +1$$

where f_i does not include local rotation around the center line of S_{13} and S_{23} (written as $R(o_{13} - o_{23})$).

(5) Determine SOC_2 and Δ_2

① According to the mechanism structure decomposition method described in Sect. 7.4.1, the second SOC is selected as

$$SOC_2\{-R(o_{13} - o_{23}) - S_{33} - P_{32}\perp R_{31}-\}, \text{ (}R_{31}\text{ is a driving pair)}$$

where, $R(o_{13} - o_{23})$ – an rotation around the center line of S_{13} and S_{23}.

② According to Eq. (6.8b) in Chap. 6, there is

$$\xi_{L_2} = \dim\{(M_{b_1} \cap M_{b_2}) \cup M_{b_3}\}$$
$$= \dim\left\{\left(\begin{bmatrix} t^3 \\ r^3 \end{bmatrix} \cap \begin{bmatrix} t^3 \\ r^3 \end{bmatrix}\right) \cup \begin{bmatrix} t^3 \\ r^3 \end{bmatrix}\right\} = \dim\left\{\begin{bmatrix} t^3 \\ r^3 \end{bmatrix}\right\} = 6$$

③ According to Eq. (7.6), there is

$$\Delta_2 = \sum_{i=1}^{m_2} f_i - I_1 - \xi_{L_1} = 6 - 1 - 6 = -1$$

(6) Determine AKCs contained in this mechanism

For the two SOCs and their constraint degree Δ_1 and Δ_2, it is easy to determine that this mechanism has one AKC according to criteria for AKC determination in Sect. 7.4.2.

$$\left.\begin{array}{ll} SOC_1\{-R_{11}\perp P_{12} - S_{13} - S_{23} - P_{22}\perp R_{21}-\}, & \Delta_1 = +1 \\ SOC_2\{-R_{31} - S_{32} - S_{33}-\} & \Delta_2 = -1 \end{array}\right\} \Rightarrow \sum_{j=1}^{2} \Delta_j = 0$$

$$\kappa = \frac{1}{2}\sum_{j=1}^{2} |\Delta_j| = 1$$

So, the parallel mechanism in Fig. 7.11 contains one AKC, i.e.

$$KC[F = 3, v = 2] = J_d[3] \oplus AKC[v = 2, \kappa = 1].$$

7.5 Determination of DOF Type Based on AKC

7.5.1 Full DOF and Its Existence Condition

(1) Definition
If position and orientation of driven links relative to the frame link are functions of all driving inputs, the mechanism has full DOF.

(2) Existence condition
When all driving pairs are located in the same AKC, the mechanism has full DOF [28–34].

For example, the parallel mechanism in Fig. 7.11 contains only one AKC. The three driving pairs are located in the same AKC. So, this parallel mechanism has full DOF. Position and orientation of the output link are functions of all the three P pairs' inputs.

7.5.2 Partial DOF and Its Existence Condition

(1) Definition
If position and orientation of some driven links relative to the frame link are functions of some driving inputs, the mechanism has partial DOF.

(2) Existence condition
When driving pairs are located in different AKCs, the mechanism has partial DOF [28–34].

For example, the parallel mechanism in Fig. 7.10a has two AKCs. The driving pairs P_2 and P_4 are located in the first AKC (Fig. 7.10d) and the driving pairs P_9 and P_{12} are located in the second AKC (Fig. 7.10e). This mechanism has partial DOF.

7.5.3 Separable DOF and Its Existence Condition

(1) Definition
If the mechanism can be separated into two or more independent sub-mechanisms, and position and orientation of the driven links in each sub-mechanism relative to the frame link are functions of only these driving inputs in this sub-mechanism, the mechanism has separable DOF.

(2) Existence condition
When driving pairs are located in different AKCs and the mechanism can be divided into two or more sub-mechanisms by separating the frame link, the mechanism has separable DOF [28–34].

For example, the mechanism in Fig. 7.12 has separable DOF.

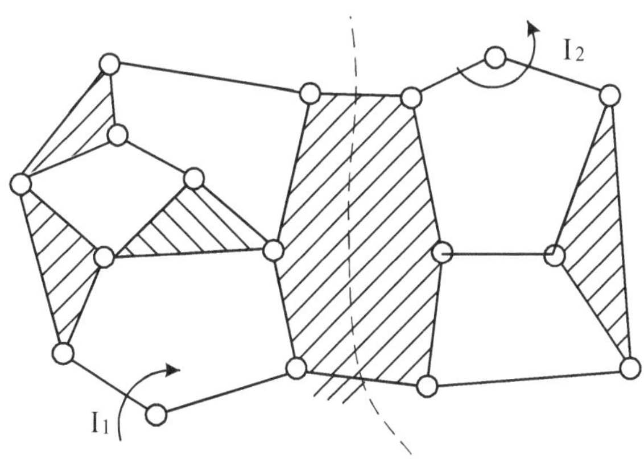

Fig. 7.12 A mechanism with separable DOF

7.6 Motion Decoupling

7.6.1 Motion Decoupling of Parallel Mechanisms

Position and orientation of the movable platform are functions of the driving inputs, i.e.

$$M_{Pa} = \begin{bmatrix} x(\theta_1, \cdots, \theta_F) & y(\theta_1, \cdots, \theta_F) & z(\theta_1, \cdots, \theta_F) \\ \alpha(\theta_1, \cdots, \theta_F) & \beta(\theta_1, \cdots, \theta_F) & \gamma(\theta_1, \cdots, \theta_F) \end{bmatrix} \tag{7.11}$$

where, M_{Pa}—position and orientation variables of the movable platform, $x(\theta_1, \cdots, \theta_F), y(\theta_1, \cdots, \theta_F), z(\theta_1, \cdots, \theta_F)$—position variables of the origin (point o′) of the moving coordinate system relative to the fixed coordinate system, $\alpha(\theta_1, \cdots, \theta_F), \beta(\theta_1, \cdots, \theta_F), \gamma(\theta_1, \cdots, \theta_F)$—orientation variables of the moving coordinate system relative to the fixed coordinate system, θ_i—the ith driving input, F—DOF of the mechanism.

According to Eq. (7.11),

(1) When each variable of the movable platform $(x, y, z, \alpha, \beta, \gamma)$ is function of all driving inputs $(\theta_1, \cdots, \theta_F)$, the mechanism has no motion decoupling property.

(2) When one (or several) variable(s) of the movable platform $(x, y, z, \alpha, \beta, \gamma)$ is (are) function(s) of some driving inputs $(\theta_1, \cdots, \theta_r)$ $(r < F)$, the mechanism is partially motion decoupled.

(3) When there is a one-to-one correspondence between the motion output variables of the movable platform and the driving inputs, the mechanism is fully motion decoupled, as shown in Eq. (7.12a) or Eq. (7.12b).

$$M_{Pa} = \begin{bmatrix} x(\theta_1) & y(\theta_2) & z(\theta_3) \\ \alpha(\theta_4) & \beta(\theta_5) & \gamma(\theta_6) \end{bmatrix} \tag{7.12a}$$

$$M_{Pa} = \begin{bmatrix} x(\theta_1) & y(\theta_1, \theta_2) & z(\theta_1 \sim \theta_3) \\ \alpha(\theta_1 \sim \theta_4) & \beta(\theta_1 \sim \theta_5) & \gamma(\theta_1 \sim \theta_6) \end{bmatrix} \tag{7.12b}$$

Motion decoupling property may simplify the kinematic and the dynamic analysis and make motion control easier.

7.6.2 Motion Decoupling Based on Partial DOF

(1) Introductory example

For the 3-loop parallel mechanism (DOF = 4) in Fig. 7.10a, $o' - x'y'z'$ is the moving coordinate system attached on the movable platform. The base point o′ (origin of the moving coordinate system) is the intersection of axes of R_6 and R_7.

According to example 7.4, this parallel mechanism contains two AKCs. AKC_1 (Fig. 7.10d) contains two driving pairs P_2 and P_4. AKC_2 (Fig. 7.10e) contains two driving pairs P_9 and P_{12}. Since these four driving pairs are arranged in two AKCs. According to types of DOF and its existence condition (Sect. 7.5.2), this parallel mechanism has partial DOF: Position of the axis of R_3 in AKC_1 is only function of inputs from driving pairs P_2 and P_4. Since R_3 and R_6 share the same axis, position of the base point o′ in the fixed coordinate system is also function of inputs from driving pairs P_2 and P_4. So motion output of the movable platform is partially decoupled.

(2) Motion decoupling principle based on partial DOF

   (a) Basic idea: For a mechanism with partial DOF, there must be a driven link whose position or orientation parameters depend on only a part of the driving inputs. When this driven link is adjacent to the movable platform and they share some common position or orientation parameters, the movable platform is partially motion decoupled [30–34].

   (b) Existence condition: The existence conditions for motion decoupling can be concluded as:

   - Based on the existence condition for partial DOF (Sect. 7.5.2), the parallel mechanism contains at least two AKCs.
   - The driven link with partial DOF must be adjacent to the movable platform and they share some common position or orientation parameters.

The motion decoupling principle based on partial DOF can be used to determine whether a parallel mechanism has motion decoupling property and can be applied to topology design of motion decoupled parallel mechanisms (refer to Chaps. 9 and 10 for detail).

7.6.3 Motion Decoupling Based on Geometrical Conditions of Assembling Branches

For a parallel mechanism containing only one AKC, the motion decoupling may be acquired by geometrical conditions of assembling branches attached between the two platforms [35–43].

For example, the 2-loops parallel mechanism in Fig. 7.13 has only one AKC. When the geometrical conditions of assembling branches are changed, the parallel mechanisms will have different motion decoupling property:

(1) If axes of two end pairs in each branch are allocated on the two platforms as shown in Fig. 7.13a, this (3T-0R) parallel mechanism has no motion decoupling property.

(2) If axes of two end pairs in each branch are allocated on the two platforms as shown in Fig. 7.13b, this (3T-0R) parallel mechanism has partial motion decoupling property. Displacements of the movable platform along y and z directions depend only on inputs of R_{21} and R_{31} and are independent of the input of R_{11}. Displacement of the movable platform along x direction depend on all the three inputs.

(3) If axes of two end pairs in each branch are allocated on the two platforms as shown in Fig. 7.13c, this (3T-0R) parallel mechanism has full motion decoupling property. Displacements of the movable platform along x, y and z directions depend only on inputs of P_{33}, P_{23} and P_{13} respectively [35–38].

Generally, the above two kinds of motion decoupling based on partial DOF or assembling conditions of branches can be combined to be used in topology design of motion decoupled PMs [35].

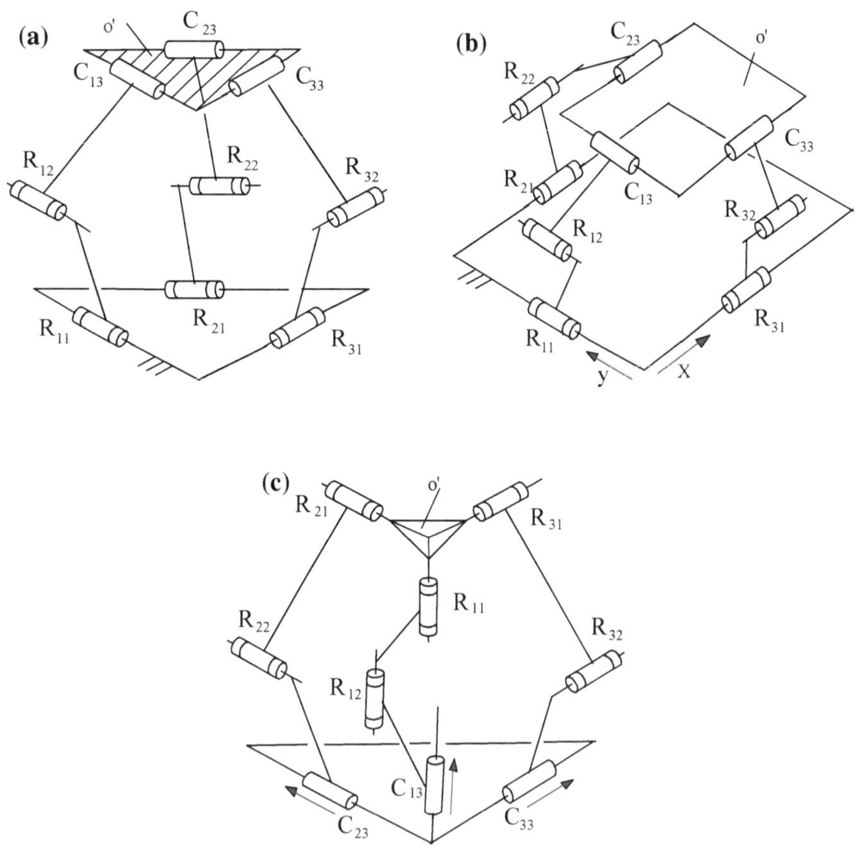

Fig. 7.13 Three (3T-0R) PMs with same branches

7.7 Applications of the Composition Principle Based on SOC Unit

According to the mechanism composition principle based on SOC unit, any mechanism can be decomposed into several AKCs and each AKC can be further decomposed into a group of ordered SOCs. Constraint degree of each SOC and coupling degree of each AKC can then be calculated. This composition principle provides a theoretical basis for unified modeling of mechanism topology, kinematics and dynamics based on SOC unit.

7.7.1 Topological Structure Synthesis Method Based on SOC Unit

(1) A general method for structure synthesis of multi-loop mechanisms based on SOC unit has been proposed [26, 27]. All types and corresponding coupling degree of planar AKCs with 1–4 independent loops have been obtained, as shown in Table 7.2 [26]:

   (a) There is only 1 type of 1-loop AKC (no. 1 in Table 7.2) and its coupling degree $\kappa = 0$;

   (b) There is only 1 type of 2-loop AKC (no. 2 in Table 7.2) and its coupling degree $\kappa = 1$;

   (c) There are 3 types of 3-loop AKC (no. 3–5 in Table 7.2) and their coupling degree $\kappa = 1$;

   (d) There are 28 types of 4-loop AKC (no. 6–33 in Table 7.2) and the coupling degree $\kappa = 1$ for AKCs from no. 6 to 29 and the coupling degree $\kappa = 2$ for AKCs from 30 to 33.
   Note: Almost all planar mechanisms with practical use contain only the first 5 types of AKCs, i.e. No. 1–No. 5 in Table 7.2.

(2) Based on the composition principle based on SOC unit, three fundamental equations for topology structure theory have been established: POC equation for serial mechanism (Eq. 4.4 in Chap. 4) [31–33], POC equation for parallel mechanism (Eq. 5.3 in Chap. 5) [31–33] and the general DOF formula (Eq. 6.8 in Chap. 6) [33]. These three equations reveal the mapping relations among topology structure, kinematic property and DOF of mechanisms.

(3) A general method for topology design of parallel mechanisms has been proposed based on these three fundamental equations [30–34]. This method is a geometrical one, which is independent on mechanism motion position and the fixed coordinate system (refer to Chaps. 8, 9, and 10).

Table 7.2 Plane AKCs ($v = 1 \sim 4$) and their coupling degrees [26, 28]

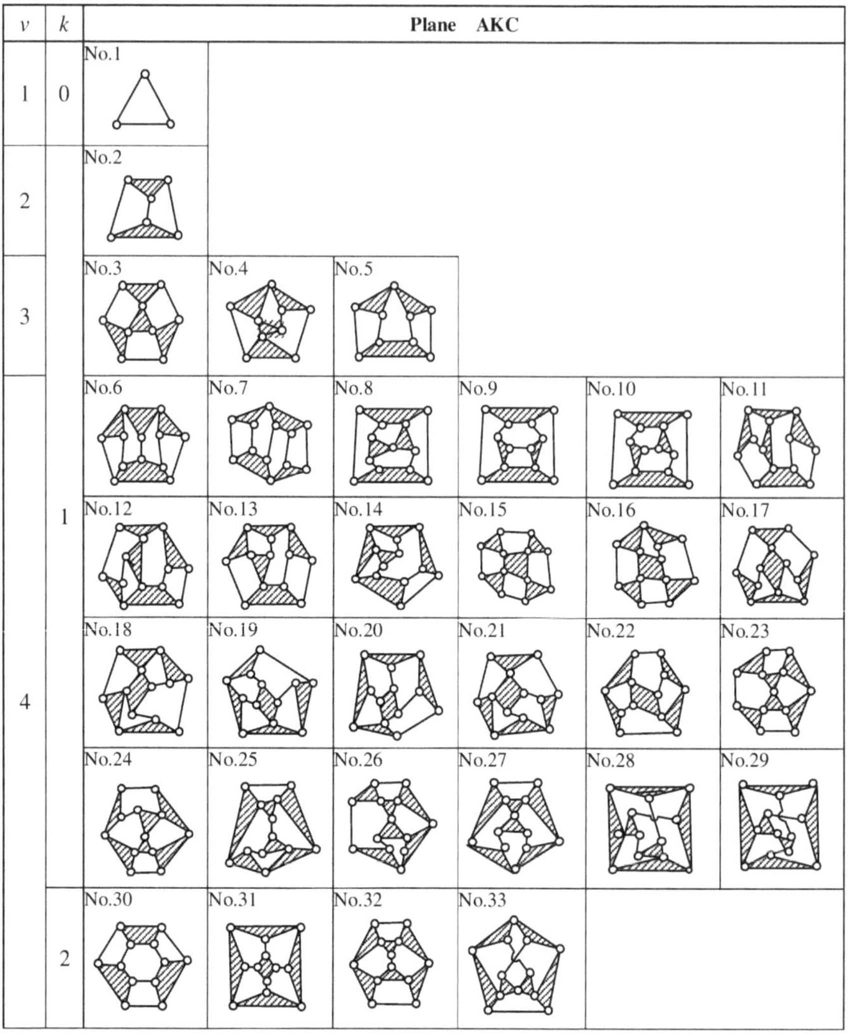

7.7.2 *Kinematic Analysis Method Based on SOC Unit*

(1) A modular kinematic analysis method for planar mechanisms based on SOC unit has been proposed [28, 44–48].

(a) Basic idea

Kinematic analysis of any mechanism can be converted to kinematic analyses of AKCs contained in this mechanism. And each AKC can be further decomposed into a group of ordered SOCs. So, kinematic analysis of an AKC can then be converted to kinematic analyses of these ordered SOC units. For each AKC, there are Δ^+ unknowns and $|\Delta^-|$ constraint equations. In another word, there are κ unknowns and κ equations. It means that dimension of the AKC's kinematic equations is κ [28, 44–47].

(b) Dimension of kinematic equations
For the three different modeling methods (link-pair unit, loop unit and SOC unit), the established kinematic equations may have the same form. But, the equation structure and generation process, especially the equation dimension (number of unknowns) may be different.

It should be noted that dimension of the planar AKC's kinematic equations established based on SOC unit is equal to its coupling degree, as shown in Table 7.3. The SOC unit method can achieve great dimension reduction at the topological structure level [44–49] and has been further developed into an elimination method which may obtain close-form solutions [46, 47].

Since all practical planar mechanisms contain only AKCs with 1–3 loops and the coupling degree is not greater than 1, all real solutions (configurations) can be obtained by one-dimension searching method.

So, the coupling degree κ can represent complexity of AKC's kinematic analysis [28, 48, 49].

(2) Based on the constraint characteristics of SOCs, an easy-to-use method for inverse displacement analysis of general 6R serial mechanism has been proposed and all its real inverse displacement solutions can be obtained by one-dimension search method [50].

(3) A modular kinematic analysis method for parallel mechanisms and multi-loop spatial mechanisms based on SOC unit has been proposed. Most parallel mechanism of practical use (such as parallel mechanisms shown in Figs. 7.6, 7.7, 7.8, 7.9, 7.10 and 7.11) contains only AKCs with $k = 1$. One-dimension searching method can be used to find all their real solutions [51–53].

Table 7.3 Dimension of kinematic equations of planar AKCs with $v = 1 \sim 4$ [28, 44–46]

Structure unit	Link-pair	Loop	SOC
Kinematic equations	• $F_g = (\theta_1, \theta_2, \ldots, \theta_f; \phi_1, \phi_2, \ldots, \phi_e; a, \alpha, d), \quad g = 1, 2, \ldots, e$ • $\sum \frac{\partial F_g}{\partial \phi_j} \dot{\phi}_j + \sum_{i=1}^{f} \frac{\partial F_g}{\partial \phi_i} \dot{\theta}_i = 0$ • $\sum [\frac{\partial F_g}{\partial \phi_j} \ddot{\phi}_j + \dot{\phi}_j \frac{d}{dt}(\frac{\partial F_g}{\partial \phi_j})] + \sum_{i=1}^{f} [\frac{\partial F_g}{\partial \theta_i} \ddot{\theta}_i + \dot{\theta}_i \frac{d}{dt}(\frac{\partial F_g}{\partial \theta_i})] = 0$		
Dimension	$e = 6\upsilon - 2m_f$	$e = 2\upsilon$	$e = \kappa = 0, 1\ or\ 2$

Note m_f—number of pairs on the frame link; $\kappa = 0$ means kinematic analysis of each SOC can be solved one by one independently

7.7.3 Dynamic Analysis Method Based on SOC Unit

(1) A modular dynamic analysis method for planar mechanisms based on SOC unit has been proposed [28, 54–56].

(a) Basic idea
Dynamic analysis of any mechanism can be converted to dynamic analyses of AKCs contained in this mechanism. And each AKC can be further decomposed into a group of ordered SOCs. So, dynamic analysis of an AKC can then be converted to dynamic analyses of these ordered SOC units. Similar to the kinematic analysis method in Sect. 7.2.2, the dynamic equations of AKC can be established based on SOC unit.

(b) Dimension of dynamic equations
For the three different modeling methods (link-pair unit, loop unit and SOC unit), the established dynamic equations may have the same form. But, the equation structure and generation process, especially the equation dimension (number of unknowns) may be different

For example, dynamic equations of planar AKCs and their dimensions are listed in Table 7.4. Dimension of the inverse dynamic equations is κ and dimension (number of unknowns) of the forward dynamic equations is $(\kappa + f)$. The method based on SOC unit can achieve great dimension reduction at the topological structure level.

So, the coupling degree κ can represent complexity of AKC's dynamic analysis [28, 48, 49].

(2) A modular dynamic analysis method for parallel mechanisms and multi-loop spatial mechanisms based on SOC unit has been proposed [55, 56].

Table 7.4 Dimension of dynamic equations of planar AKCs with $v = 1 \sim 4$ [28, 54]

Structure unit		Link-pair	Loop	SOC
Dynamic equations		• Inverse: $R_g(F_1, \ldots, F_n; M_1, \ldots, M_n; R_1, \ldots, R_e) = 0, \quad g = 1, 2, \ldots, e$ • Forward: $[A]^T_{e\times(f+e)}[R]_{e\times 1} + [H]_{(f+e)\times(f+e)}[\ddot{q}]_{(f+e)\times 1} + [P(\dot{q})]_{(f+e)\times 1}[P]_{(f+e)\times 1} = 0$ $[\ddot{q}(e)] = [L][\ddot{q}(f)] + [\dot{L}][\dot{q}(f)]$		
Dimension	Inverse	$e = 6\upsilon - 2m_f$	$e = 2\upsilon$	$e = \kappa = 0, 1\ or\ 2$
	Forward	$e = 6\upsilon - 2m_f + f$	$e = 2\upsilon + f$	$e = \kappa + f, \quad \kappa = 0, 1 \quad or\ 2$

Note m_f is number of pairs on the frame link, f is the DOF, $\kappa = 0$ means dynamic analysis of each SOC can be solved one by one independently

7.8 Summary

(1) The mechanism composition principle based on SOC unit is proposed. Constraint degree of SOC is defined and is used to describe constraint characteristics of SOC unit to DOF of the mechanism.

(2) A general method for decomposing AKC into ordered SOCs is proposed. The coupling degree of AKC is defined. So, kinematic (dynamic) analysis of an AKC is converted to kinematic (dynamic) analysis of ordered SOCs and dimension (number of unknowns) of the kinematic analysis equations is just its coupling degree. Therefore, coupling degree is a topological invariant and it may represent the AKC's complexity.

(3) A general method for decomposing mechanism into ordered SOCs is proposed. Criteria for AKC determination is defined and is used to determine types and number of AKCs contained in the mechanism.

(4) A general method for determination of DOF types is proposed based on the AKCs contained in the mechanism.

(5) There are two kinds of motion decoupling for PMs: based on partial DOF of a PM and based on geometrical conditions of assembling branches. A general method for motion decoupling design based on partial DOF is proposed.

(6) According to the mechanism composition principle based on SOC unit, a mechanism can be decomposed into a group of ordered SOCs. Constraint degree of SOC and coupling degree of AKC can be obtained. A systematic theory based on SOC unit for mechanism topology, kinematics and dynamics could be established.

References

1. Assur LV (1913) Investigation of Plane Hinged Mechanisms with Lower Pairs from the Point of View of Their Structure and Classification (in Russian): Part I, II. Bull. Petrograd Polytech Inst 20 :329–386. Bull Petrograd Polytech Inst 21 (1914) 187–283, (vols. 21–23, 1914–1916)
2. Dobrovolskii VV (1939) Main principles of rational classification. AS USSR
3. Verho A (1973) An extension of the concept of the group. Mech Mach Theor 8(2):249–256. doi:10.1016/0094-114X(73)90059-1
4. Manolescu NI (1979) A unified method for the formation of all planar jointed kinematic chains and Baranov trusses. Environ Plann B 6(4):447–454
5. Galletti C (1986) A note on modular approaches to planar linkage kinematic analysis. Mech Mach Theor 21(5):385–391
6. Fanghella P (1988) Kinematics of spatial linkages by group algebra: a structure-based approach. Mech Mach Theor 23(3):171–183. doi:10.1016/0094-114X(88)90102-4
7. Ceresole E, Fanghella P, Galletti C (1996) Assur's groups, AKCs, basic trusses, SOCs, etc.: modular kinematics of planar linkages. In: Proceedings of the 1996 ASME design engineering technical conference 96-DETC/MECH-1027
8. Mruthyunjaya TS (2003) Kinematic structure of mechanisms revisited. Mech Mach Theor 38:279–320

9. Servatius B, Shai O, Whiteley W (2010) Combinatorial Characterization of the Assur graphs from engineering. Eur J Comb 31(4):1091–1104
10. Shai O, Sljoka A, Whiteley W (2013) Directed graphs, decompositions, and spatial rigidity. Discr Appl Math 161:3028–3047
11. Reuleaux F (1876) Theoretische kinematic. Fridrich vieweg, Braunschweig, Germany, 1875 (Kennedy ABW, (1876) The kinematics of machinery. Reprinted Dover 1963)
12. Franke R (1951) Vom Aufbau der Getriebe, Vol. 1", Beuthvertrieb, Berlin, 1943, 2nd edition, VDI, Verlag (1958). R ~ Franke, Vom Aufbau der Getriebe, Vol. 1, 2nd edition VDI Verlag (1958).R ~ Franke, Getriebe
13. Beyer R., 1963, " The Kinematic Synthesis of Mechanisms", McGraw-hill Book Comp
14. Hain K (1967) Applied kinematics, second edn. McGraw-Hill, New York
15. Woo LS (1967) Type synthesis of plane linkages. ASME J Eng Ind B 89:159–172
16. Tischler C, Samuel A, Hunt KH (1995) Kinematic chains for robot hands—I. Orderly number-synthesis. Mech Mach Theor 30(8):1193–1215. Retrieved from http://www.sciencedirect.com/science/article/pii/0094114X9500043X
17. Wittenburg J (1977) Dynamics of Systems of Rigid Bodies. Teubner, Stuttgart
18. Dobrjanskyj L, Freudenstein F (1967) Some applications of graph theory to the structural analysis of mechanisms. J Eng Ind 89:153–158
19. Paul B (1979) Kinematics and dynamics of planar machinery. Prentice-Hall Inc, New Jersey
20. Kinzel GL et al (1984) The analysis of planar linkages using a modular approach. Mech Mach Theor 19:165–172
21. Sohn WJ, Freudenstein F (1989) An application of dual graphs to the automatic generation of the kinematic structure of mechanisms. ASME J Mech Trans Auto Design 111(4):494–497
22. Tuttle ER et al (1989) Enumeration of basic kinematic chains using the theory of finite groups. Trans ASME J Mech Trans Auto Design 111(4):498–503
23. Waldron KJ, Sreenivasan SV (1996) A study of the solvability of the position problem for multi-circuit mechanisms by way of example of the double butterfly linkage. ASME J Mech Design 118:390–395
24. Tsai L-W (1999) Robot analysis: the mechanics of serial and parallel manipulators. John Wiley, New York
25. McCarthy JM, Soh GS (2011) Geometric design of linkages. Springer, New York
26. Yang T-L, Yao F-H (1988) The topological characteristics and automatic generation of structural analysis and synthesis of plane mechanisms, part 1-theory, part 2-application. In: Proceedings of ASME mechanisms conference, vol (1). pp 179–190
27. Yang T-L, Yao F-H (1992) The topological characteristics and automatic generation of structural analysis and synthesis of spatial mechanisms, part 1-topological characteristics of mechanical network; part 2-automatic generation of structure types of kinematic chains. In: Proceedings of ASME mechanisms conference DE-47. pp 179–190
28. Yang T-L (1996) Basic theory of mechanical system: structure, kinematic and dynamic. China machine Press, Beijing
29. Yang T-L, Yao F-H, Zhang M (1998) A comparative study on some modular approaches for analysis and synthesis of planar linkages: part I—modular structural analysis and modular kinematic analysis. In: Proceedings of the ASME mechanisms conference Atlanta, DETC98/MECH-5920
30. Yang T-L, Liu A-X, Shen H-P et al (2015) Composition principle based on SOC unit and coupling degree of BKC for general spatial mechanisms. The 14th IFToMM World Congress, Taipei, Taiwan, 25–30 October 2015. doi: 10.6567/IFToMM.14TH.WC.OS13.135
31. Jin Q, Yang T-L (2004) Theory for topology synthesis of parallel manipulators and its application to three dimension translation parallel manipulators. ASME J Mech Des 126:625–639
32. Yang T-L (2004) Theory of topological structure for robot mechanisms. China Machine Press, Beijing
33. Yang T-L, Sun D-J (2012) A general DOF formula for parallel mechanisms and multi-loop spatial mechanisms. ASME J Mech Robot 4(1):011001

34. Yang T-L, Liu A-X, Luo Y-F et al (2012) Theory and application of robot mechanism topology. China Science Press, Beijing
35. Jin Q, Yang T-L (2004) Synthesis and analysis of a group of 3-degree-of-freedom partially decoupled parallel manipulators. ASME J Mech Des 126:301–306
36. Carricato Marco (2005) Fully isotropic four-degrees-of-freedom parallel mechanisms for Schoenflies motion. Int J Robot Res 24(5):397–414
37. Sébastien B, Arakelian V, Guégan S (2008) Design and prototyping of a partially decoupled 4-DOF 3T1R parallel manipulator with high-load carrying capacity. J Mech Design 130:122303
38. Lee Chung-Ching, Hervé Jacques M (2009) Uncoupled actuation of overconstrained 3T-1R hybrid parallel manipulators. Robotica 27:103–117
39. Carricato M (2009) Decoupled and homokinetic transmission of rotational motion via constant-velocity joints in closed-chain orientational manipulators. J Mech Robot 1:041008
40. Glazunov Victor (2010) Design of decoupled parallel manipulators by means of the theory of screws. Mech Mach Theor 45:239–250
41. Carricato Marco, Parenti-Castelli Vincenzo (2004) A novel fully decoupled two degrees of freedom parallel wrist. Int J Robot Res 23(6):661–667
42. Kong X, Gosselin CM (2002) Type synthesis of linear translational parallel manipulators. Adv Robot Kinematics—Theor Appl 411–420
43. Li Weimin, Gao Feng, Zhang Jianjun (2005) R-CUBE, a decoupled parallel manipulator only with revolute joints. Mech Mach Theor 40:467–473
44. Shen H-P, Yang T-L (2000) A new method and automatic generation for kinematic analysis of complex planar linkages based on the ordered SOC. In: Proceedings of ASME mechanisms conference, vol 70. pp 493–500
45. Shen H-P, Ting K-L, Yang T-L (2000) Configuration analysis of complex multiloop linkages and manipulators. Mech Mach Theor 35(3):353–362
46. Hang L-B, Jin Q, Jin Wu, Yang T-L (2000) A general study of the number of assembly configurations for multi-circuit planar linkages. J Southeast Univ (English Ed.) 16(1):46–51
47. Nicolas R, Federico T (2012) On closed-form solutions to the position analysis of Baranov trusses. Mech Mach Theor 50:179–196
48. Hahn E, Shai O (2016) A single universal construction rule for the structural synthesis of mechanisms. In: Proceedings of the ASME international design engineering technical conference IDETC/CIE 2016-59133
49. Hahn E, Shai O (2016) Construction of Baranov trusses using a single universal construction rule. In: Proceedings of the ASME international design engineering technical conference IDETC/CIE 2016-59134
50. Shi Z-X, Luo Y-F, Hang L-B, Yang T-L (2007) A simple method for inverse kinematic analysis of the general 6R serial robot. ASME J Mech Design 129(8):793–798
51. Feng Z-Y, Zhang C, Yang T-L (2006) Direct displacement solution of 4-DOF spatial parallel mechanism based on ordered single-open-chain. Chinese J Mech Eng 42(7):35–38
52. Shi Z-X, Luo Y-F, Yang T-L (2006) Modular method for kinematic analysis of parallel manipulators based on ordered SOCs. In: Proceedings of the ASME 31th mechanisms and robots confernce DETC2006-99089
53. Huiping S, Guowei S, Jiaming D, Ting-li Y (2017) A novel 3T1R parallel robot 2PaRSS: design and kinematics. In: Proceedings of the ASME 2017 international design engineering technical conferences, Ohio, DETC2017-67265
54. Zhang J-Q, Yang T-L (1994) A new method and automatic generation for dynamic analysis of complex planar mechanisms based on the SOC. In: Proceedings of 1994 ASME mechanisms conference, vol 71. pp 215–220
55. Yang T-L, Li H-L, Luo Y-F (1991) On the structure of dynamic equation of any mechanical system. Chinese J Mech Eng 27(4):1–15
56. Yang T-L, Yao F-H, Zhang M (1998) A comparative study on some modular approaches for analysis and synthesis of planar linkages: Part II — Modular Dynamic Analysis, Modular Structural Synthesis and Modular Kinematic Synthesis," Proc. of the ASME Mechanisms Conf., DETC98/MECH-6058

Chapter 8
General Method for Structure Synthesis of Serial Mechanisms

8.1 Introduction

Structure synthesis of serial mechanisms can be used not only in topology design of serial robot mechanisms [1–5], but in structure synthesis for branches of parallel mechanisms [6–10].

The fundamental requirements of serial mechanism structure synthesis are (a) the serial mechanism has the specified DOF and (b) the end link has the prescribed POC set. Structure synthesis involves: type and number of kinematic pairs, connection relations among links and geometric constraint types among axes of kinematic pairs (parallel, coaxial, perpendicular, intersecting at a common point, coplanar, skew, etc.). These three key elements can be combined in so many different ways. So, the structure synthesis is a very complex issue.

There are four main types of method for structure synthesis of branches (i.e. serial mechanisms) of parallel mechanism:

(1) Method based on screw theory [1–3]

Basic idea of this method: Constraint screw system of each branch is obtained by decomposing constraint screw system of the parallel mechanism's movable platform. Then motion screw system is obtained by reciprocal product. Select the type of kinematic pair through linear combination of each branch's motion screw system. Then arrange the order to connect kinematic pairs and obtain topological structure of branches. This method depends on motion position of the mechanism.

(2) Method based on displacement subgroup/submanifold [4–6]

Basic idea of this method: The branch is designed based on certain types of displacement subgroups. But so far, no systematic design method has been established [6].

(3) Method based on theory of linear transformations and evolutionary morphology [7]

T.-L. Yang et al., *Topology Design of Robot Mechanisms*,
https://doi.org/10.1007/978-981-10-5532-4_8

Basic idea of this method: Type and number of kinematic pairs, geometric constraint types of pair axes, etc. are determined based on morphological operations (i.e. the combination, the mutation, the migration and the selection). A library of branches with connectivity 2 to 6 is established. This method depends on motion position of a mechanism.

(4) Method based on POC equation for serial mechanisms [8–15]

Basic idea of this method: Structure types of serial mechanism containing only R and P pairs are determined based on POC sets of kinematic pairs (Eqs. 3.1–3.3 in Chap. 3), POC sets of sub-SOCs of geometric constraint types of axes (Table 4.1 in Chap. 4) and POC equation for serial mechanisms (Eq. 4.4 in Chap. 4). Then structure types of serial mechanisms containing multi-DOF pairs (such as C, U, S and E) are obtained through expansion. In addition, a simple method for structure synthesis of single-loop mechanisms is proposed [13–17]. This method is independent of motion position of mechanism and the fixed coordinate system (i.e. it is not necessary to establish the fixed coordinate system).

8.2 Method for Structure Synthesis of Serial Mechanisms

8.2.1 Basic Requirements of Structure Synthesis

(1) POC set of the serial mechanism (M_S). This POC set should not contain non-independent element.

(2) DOF of the serial mechanism. There are the following two cases:

  (a) DOF is equal to dimension (number of independent elements) of the POC set, i.e. $DOF = \dim\{M_S\}$.

  (b) DOF is greater than dimension of the POC set, i.e. $DOF > \dim\{M_S\}$.

8.2.2 General Procedure for Structure Synthesis of Serial Mechanisms

The general procedure for structure synthesis of serial mechanisms is shown in Fig. 8.1.

8.2.3 Main Steps of Structure Synthesis

Step 1 POC set and DOF of the serial mechanism are specified.

Step 2 Determine pair combination schemes
According to POC equation for serial mechanisms (Eq. 4.4 in Chap. 4), the number of pairs in the mechanism containing only R and P pairs shall comply with the following requirements:

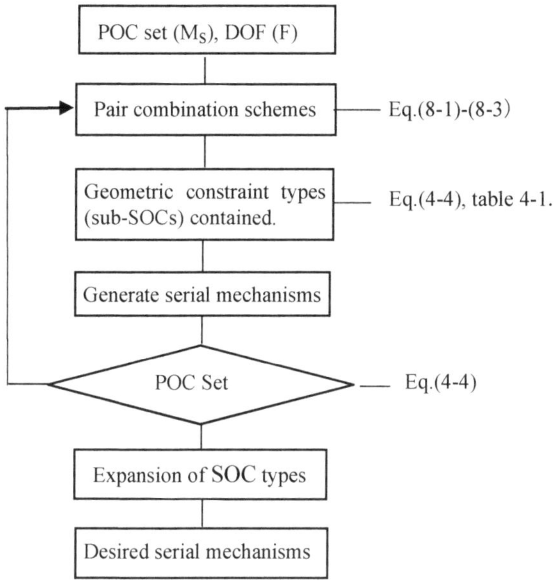

Fig. 8.1 Flowchart for structure synthesis of serial mechanisms

(1) DOF of the mechanism

$$F = m_R + m_P \tag{8.1}$$

where, F—DOF of the mechanism, m_R—number of R pairs, m_P—number of P pairs.

(2) Number of independent rotational elements.

$$m_R \geq \dim\{M_S(r)\} \tag{8.2}$$

where, $\dim\{M_S(r)\}$—number of independent rotational elements in the POC set.

(3) Number of independent translational elements.

$$m_P \leq \dim\{M_S(t)\} \tag{8.3}$$

where, $\dim\{M_S(t)\}$—number of independent translational elements in the POC set.

Obviously, there may be many pair combination schemes which satisfy Eqs. (8.1)–(8.3).

Step 3 Determine sub-SOCs contained in the serial mechanism
Determine sub-SOCs contained in the serial mechanism based on POC sets of kinematic pairs (Eqs. 3.1–3.3 in Chap. 3), POC sets of sub-SOCs of geometric constraint types of axes (Table 4.1 in Chap. 4) and POC equation for serial mechanisms (Eq. 4.4 in Chap. 4).

This is an essential step in the structure synthesis. Using sub-SOCs instead of individual kinematic pairs may simplify the calculation and even the whole structure synthesis process.

Step 4 Generate structure types of the serial mechanism.
Connect the sub-SOCs determined in step 3 in series and generate structure types of the serial mechanism.

Step 5 Check POC set of the serial mechanism's end link.
For the structure types obtained in step 4, check whether the POC set complies with design requirement based on POC equation for serial mechanism (Eq. 4.4).

Step 6 Obtain more structure types through expansion.
For the obtained structure types containing only R and P pairs, more structure types can be obtained using the following methods while the POC set is kept unchanged.

(1) Change the connection order of kinematic pairs
For example, the following four types of SOC branches can be obtained by changing the connection order of kinematic pairs:

① $SOC\{-R \parallel R - P - P-\}$;

② $SOC\{-R(-P) \parallel R - P-\}$, axes of the two R pairs are parallel.

③ $SOC\{-R(-P - P) \parallel R-\}$, axes of the two R pairs are parallel.

④ $SOC\{-P - R \parallel R - P-\}$.

(2) Combine some R pairs and P pairs into C pairs or U pairs.
For example, for $SOC\{-R \parallel R \parallel R - P-\}$, if the R pair and the P pair are combined into a C pair, $SOC\{-R \parallel R \parallel C-\}$ can be obtained. For $SOC\{-R \parallel R \parallel R - R \parallel R-\}$, if the third R pair and the fourth R pair are combined into a U pair, $SOC\{-R \parallel R \parallel \overbrace{R \perp R} \parallel R-\}$ can be obtained ($\overbrace{R \perp R}$ is the symbolic representation of U pair) (refer to Chap. 2).

(3) Replace R pairs with an S pair.

For example, $SOC\{-\overbrace{R_1 R_2 R_3}-\}$ in which axes of the three R pairs intersect at a common point can be replaced by a S pair.

8.2.4 Examples

Example 8.1 Structure synthesis of serial mechanism with DOF = $\dim\{M_S\}$.

Step 1 Basic requirements

(1) POC set: $M_S = \begin{bmatrix} t^2 \\ r^3 \end{bmatrix}$, i.e. $\dim\{M_S(r)\} = 3$, $\dim\{M_S(t)\} = 2$. Base point o′ is on axis of the end pair or is the common interesting point of several R pairs.

(2) DOF: F = $\dim\{M_S\}$ = 5.

Step 2 Determine pair combination schemes
Since $\dim\{M_S(r)\} = 3$ and $\dim\{M_S(t)\} = 2$, there are the following three pair combination schemes according to Eqs. (8.1)–(8.3):

Case 1: 5R;

Case 2: 4R1P;

Case 3: 3R2P.

Step 3 Determine sub-SOCs possibly contained in the serial mechanism

Case 1: 5R

(1) In order to achieve $\dim\{M_S(r)\} = 3$, according to POC equation for serial mechanisms (Eq. 4.4 in Chap. 4), this serial mechanism should contain at least three R pairs whose axes are not parallel.

(2) In order to achieve $\dim\{M_S(t)\} = 2$, according to Table 4.1 in Chap. 4 (POC sets of sub-SOCs), this serial mechanism should possibly contain the following sub-SOCs:

(a) $SOC\{-R \parallel R-\}$ (No. 1 in Table 4.1).

(b) $SOC\{-R \parallel R \parallel R-\}$ (No. 2 in Table 4.1).

(c) $SOC\{-\widehat{RR}-\}$ (No. 5 in Table 4.1).

(d) $SOC\{-\widehat{RRR}-\}$ (No. 6 in Table 4.1).

(3) Since several R pairs have already achieved $\dim\{M_S(t)\} = 2$, the other R pairs should not generate derivative translations of the base point o′ on the end link. So, according to Table 4.1 (POC sets of sub-SOCs), this serial mechanism should contain the following sub-SOCs:

(a) $SOC\{-\widehat{RR}-\}$ (No. 5^* in Table 4.1), arranged at end of the serial mechanism.

(b) $SOC\{-\widehat{RRR}-\}$ (No. 6^* in Table 4.1), arranged at end of the serial mechanism.

Case 2: 4R1P

(1) In order to achieve $\dim\{M_S(r)\} = 3$, according to POC equation for serial mechanisms (Eq. 4.4), this serial mechanism should contain at least three R pairs whose axes are not parallel.

(2) In order to achieve $\dim\{M_S(t)\} = 2$, according to Table 4.1 (POC sets of sub-SOCs), this serial mechanism should possibly contain the following sub-SOCs:

 (a) $SOC\{-R\perp P-\}$ (No. 1 in Table 4.1).

 (b) $SOC\{-R \parallel R\perp P-\}$ (No. 2 in Table 4.1).

(3) Since several R pairs and P pairs have already achieved $\dim\{M_S(t)\} = 2$, the other R pairs should not generate derivative translations of the base point o′ on the end link. So, according to Table 4.1 (POC sets of sub-SOCs), this serial mechanism should contain the following sub-SOCs:

 (a) $SOC\{-\widehat{RR}-\}$ (No. 5[*] in Table 4.1), arranged at end of the serial mechanism.

 (b) $SOC\{-\widehat{RRR}-\}$ (No. 6[*] in Table 4.1), arranged at end of the serial mechanism.

Case 3: 3R2P

(1) In order to achieve $\dim\{M_S(r)\} = 3$, according to POC equation for serial mechanisms (Eq. 4.4), this serial mechanism should contain at least three R pairs whose axes are not parallel.

(2) In order to achieve $\dim\{M_S(t)\} = 2$, according to Table 4.1 (POC sets of sub-SOCs), this serial mechanism should possibly contain the following sub-SOCs:

 (a) $SOC\{-\diamondsuit(P,P)-\}$ (No. 3 in Table 4.1).

 (b) $SOC\{-P\perp R\perp P-\}$ (No. 2 in Table 4.1).

(3) Since several R pairs and P pairs have already achieved $\dim\{M_S(t)\} = 2$, the other R pairs should not generate derivative translations of the base point o′ on the end link. So, according to Table 4.1 (POC sets of sub-SOCs), this serial mechanism should contain the following sub-SOCs:

 (a) $SOC\{-\widehat{RR}-\}$ (No. 5[*] in Table 4.1), arranged at end of the serial mechanism.

 (b) $SOC\{-\widehat{RRR}-\}$ (No. 6[*] in Table 4.1), arranged at end of the serial mechanism.

Step 4 Generate structure types of the serial mechanism.
Connect the sub-SOCs determined in step 3 in series and generate structure types of the serial mechanism.

Case 1: 5R

(1) $SOC\{-R \parallel R - \widehat{RRR} -\}$.

(2) $SOC\{-R \parallel R \parallel R - \widehat{RR} -\}$.

(3) $SOC\{- \widehat{RR} - \widehat{RRR} -\}$.

(4) $SOC\{- \widehat{RRR} - \widehat{RR} -\}$.

Case 2: 4R1P

(1) $SOC\{-R \perp P - \widehat{RRR} -\}$.

(2) $SOC\{-R(\perp P) \parallel R - \widehat{RR} -\}$.

Case 3: 3R2P

(1) $SOC\{-P - P - \widehat{RRR} -\}$.

(2) $SOC\{-P \perp R \perp P - \widehat{RR} -\}$.

Step 5 Check POC set of the serial mechanism
According to POC equation for serial mechanism (Eq. 4.4), it is easy to determine that POC sets of the eight serial mechanisms obtained in step 4 comply with the design requirements.

Step 6 Obtain more structure types through expansion.

For the above eight structure types, replace sub-$SOC\{- \widehat{RRR} -\}$ with an S pair and obtain the following four new structure types:

(1) $SOC\{- \widehat{RR} -S-\}$.

(2) $SOC\{-R \parallel R - S-\}$.

(3) $SOC\{-R \perp P - S-\}$.

(4) $SOC\{-P - P - S-\}$.

Example 8.2 Structure synthesis of serial mechanism with DOF = $\dim\{M_S\} + 1$

Step 1 Basic requirements

(1) POC set: $M_S = \begin{bmatrix} t^2 \\ r^3 \end{bmatrix}$, i.e.$\dim\{M_S(r)\} = 3, \ \dim\{M_S(t)\} = 2$. Base point o′ is on axis of the end pair or is the common interesting point of several R pairs.

(2) DOF: F = $\dim\{M_S\} + 1 = 6$.

Step 2 Determine pair combination schemes

Since $\dim\{M_S(r)\} = 3$ and $\dim\{M_S(t)\} = 2$, there are the following three pair combination schemes according to Eqs. (8.1)–(8.3):

Case 1: 6R;

Case 2: 5R1P;

Case 3: 4R2P.

Step 3 Determine sub-SOCs possibly contained in the serial mechanism

Case 1: 6R

(1) In order to achieve $\dim\{M_S(r)\} = 3$, according to POC equation for serial mechanisms (Eq. 4.4 in Chap. 4), this serial mechanism should contain at least three R pairs whose axes are not parallel.

(2) In order to achieve $\dim\{M_S(t)\} = 2$, according to Table 4.1 in Chap. 4 (POC sets of sub-SOCs), this serial mechanism should possibly contain the following sub-SOCs:

(a) $SOC\{-R \parallel R-\}$ (No. 1 in Table 4.1).

(b) $SOC\{-R \parallel R \parallel R-\}$ (No. 2 in Table 4.1).

(c) $SOC\{-\overbrace{RR}-\}$ (No. 5 in Table 4.1).

(d) $SOC\{-\overbrace{RRR}-\}$ (No. 6 in Table 4.1).

(3) Since several R pairs have already achieved $\dim\{M_S(t)\} = 2$, the other R pairs should not generate derivative translations of the base point o′ on the end link. So, according to Table 4.1 (POC sets of sub-SOCs), this serial mechanism should contain the following sub-SOCs:

(a) $SOC\{-\overbrace{RR}-\}$ (No. 5* in Table 4.1), arranged at end of the serial mechanism.

(b) $SOC\{-\overbrace{RRR}-\}$ (No. 6* in Table 4.1), arranged at end of the serial mechanism.

Case 2: 5R1P

(1) In order to achieve $\dim\{M_S(r)\} = 3$, according to POC equation for serial mechanisms (Eq. 4.4), this serial mechanism should contain at least three R pairs whose axes are not parallel.

(2) In order to achieve $\dim\{M_S(t)\} = 2$, according to Table 4.1 (POC sets of sub-SOCs), this serial mechanism should possibly contain the following sub-SOCs:

(a) $SOC\{-R\perp P-\}$ (No. 1 in Table 4.1).

(b) $SOC\{-R \parallel R\perp P-\}$ (No. 2 in Table 4.1).

(3) Since several R pairs and P pairs have already achieved $\dim\{M_S(t)\} = 2$, the other R pairs should not generate derivative translations of the base point o′ on the end link. So, according to Table 4.1 (POC sets of sub-SOCs), this serial mechanism should contain the following sub-SOCs:

(a) $SOC\{-\widehat{RR}-\}$ (No. 5* in Table 4.1), arranged at end of the serial mechanism.

(b) $SOC\{-\widehat{RRR}-\}$ (No. 6* in Table 4.1), arranged at end of the serial mechanism.

Case 3: 4R2P

(1) In order to achieve $\dim\{M_S(r)\} = 3$, according to POC equation for serial mechanisms. (Eq. 4.4 in Chap. 4), this serial mechanism should contain at least three R pairs whose axes are not parallel.

(2) In order to achieve $\dim\{M_S(t)\} = 2$, according to Table 4.1 in Chap. 4 (POC sets of sub-SOCs), this serial mechanism should possibly contain the following sub-SOCs:

(a) $SOC\{-\Diamond(P,P)-\}$ (No. 3 in Table 4.1).

(b) $SOC\{-P\perp R\perp P-\}$ (No. 2 in Table 4.1).

(3) Since several R pairs and P pairs have already achieved $\dim\{M_S(t)\} = 2$, the other R pairs should not generate derivative translations of the base point o′ on the end link. So, according to Table 4.1 (POC sets of sub-SOCs), this serial mechanism should contain the following sub-SOCs:

(a) $SOC\{-\widehat{RR}-\}$ (No. 5* in Table 4.1), arranged at end of the serial mechanism.

(b) $SOC\{-\widehat{RRR}-\}$ (No. 6* in Table 4.1), arranged at end of the serial mechanism.

Step 4 Generate structure types of the serial mechanism.
Connect the sub-SOCs determined in step 3 in series and generate structure types of the serial mechanism.

Case 1: 6R

(1) $SOC\{-\widehat{RRR}-\widehat{RRR}-\}$.

(2) $SOC\{-R\parallel R\parallel R-\widehat{RRR}-\}$.

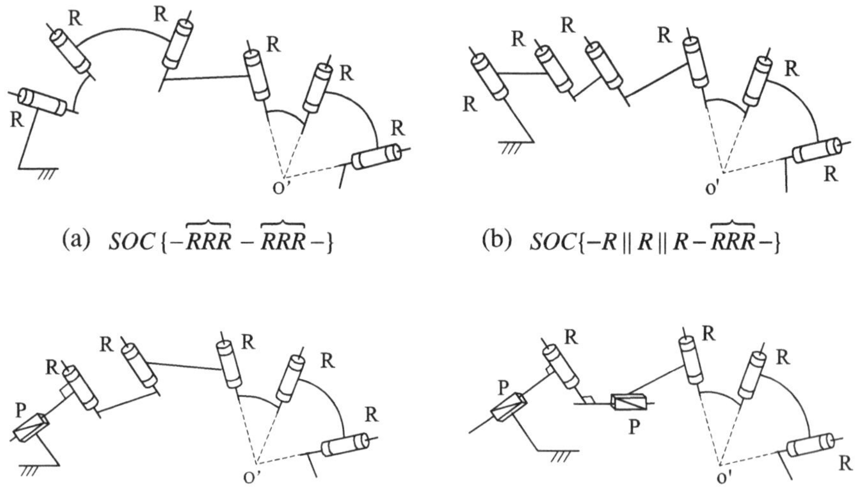

(a) $SOC\{-\widehat{RRR}-\widehat{RRR}-\}$ (b) $SOC\{-R \parallel R \parallel R-\widehat{RRR}-\}$

(c) $SOC\{-P \perp R \parallel R-\widehat{RRR}-\}$ (d) $SOC\{-P \perp R \perp P-\widehat{RRR}-\}$

Fig. 8.2 Structure types of the (2T-3R) serial mechanisms with DOF=6

Case 2: 5R1P

(1) $SOC\{-R \parallel R \perp P-\widehat{RRR}-\}$.

Case 3: 4R2P

(1) $SOC\{-P \perp R \perp P-\widehat{RRR}-\}$.

Step 5 Check POC set of the serial mechanism
According to POC equation for serial mechanism (Eq. 4.4), it is easy to determine that POC sets of the four serial mechanisms obtained in step 4 comply with the design requirements.

Step 6 Obtain more structure types through expansion. (Omitted here.)

Step 7 Draw schematic diagrams of the obtained serial mechanisms
The obtained four structure types of the serial mechanism are shown in Fig. 8.2.

8.3 Structure Types of Serial Mechanisms

8.3.1 Structure Types Containing Only R Pairs and P Pairs

Using the method for structure synthesis of serial mechanisms described in Sect. 8.2, 28 structure types of serial mechanisms containing only R and P pairs with DOF = dim$\{M_S\}$ = 2–5 and 15 structure types of serial mechanisms containing only R and P pairs with DOF = dim$\{M_S\}$ + 1 are obtained, as shown in Table 8.1 [8–12, 17].

Table 8.1 Structure types of serial mechanisms containing only R and P pairs [8–17]

dim{M$_S$}	M_s	$DOF = \dim\{M_S\}$	$DOF = \dim\{M_S\} + 1$
2	$\begin{bmatrix} t^2 \\ r^0 \end{bmatrix}$	$SOC\{-P-P-\}$	$SOC\{\diamondsuit(-P-P-P-)\}$
	$\begin{bmatrix} t^1 \\ r^1 \end{bmatrix}$	$SOC\{-P-R-\}$	
	$\begin{bmatrix} t^0 \\ r^2 \end{bmatrix}$	$SOC\{-\widehat{RR}-\}$	
3	$\begin{bmatrix} t^3 \\ r^0 \end{bmatrix}$	$SOC\{-P-P-P-\}$	$SOC\{-P-P-P-P-\}$
	$\begin{bmatrix} t^2 \\ r^1 \end{bmatrix}$	$SOC\{-R \parallel R \parallel R-\}$ $SOC\{-R \parallel R \perp P-\}$ $SOC\{-P \perp R \perp P-\}$	$SOC\{-R \parallel R \parallel R \parallel R-\}$ $SOC\{-R \parallel R \parallel R \perp P-\}$ $SOC\{-R \parallel R \parallel R \perp P-\}$
	$\begin{bmatrix} t^1 \\ r^2 \end{bmatrix}$	$SOC\{-P-\widehat{RR}-\}$	
	$\begin{bmatrix} t^0 \\ r^3 \end{bmatrix}$	$SOC\{-\widehat{RRR}-\}$	$SOC\{-\widehat{RRRR}-\}$
4	$\begin{bmatrix} t^3 \\ r^1 \end{bmatrix}$	$SOC\{-R \parallel R \parallel R-P-\}$ $SOC\{-R \parallel R-P-P-\}$ $SOC\{-R-P-P-P-\}$	$SOC\{-R \parallel R \parallel R-P-P-\}$ $SOC\{-R \parallel R-P-P-P-\}$
	$\begin{bmatrix} t^2 \\ r^2 \end{bmatrix}$	$SOC\{-R \parallel R \parallel \widehat{RR}-\}$ $SOC\{-P \perp R \parallel \widehat{RR}-\}$ $SOC\{-P-P-\widehat{RR}-\}$	
	$\begin{bmatrix} t^1 \\ r^3 \end{bmatrix}$	$SOC\{-R-\widehat{RRR}-\}$ $SOC\{-P-\widehat{RRR}-\}$	

(continued)

Table 8.1 (continued)

dim{M$_S$}	M_s	$DOF = \dim\{M_S\}$	$DOF = \dim\{M_S\} + 1$
5	$\begin{bmatrix} t^3 \\ r^2(\parallel \Diamond(R, R^*)) \end{bmatrix}$	$SOC\{-R \parallel R \parallel R - R^* \parallel R^* -\}$ $SOC\{-R \parallel R \parallel R - R^* - P-\}$ $SOC\{-R \parallel R - R^* \parallel R^* - P-\}$ $SOC\{-R \parallel R - R^* - P - P-\}$ $SOC\{-P - P - P - R - R^* -\}$	$SOC\{-R \parallel R \parallel R - R^* \parallel R^* \parallel R^* -\}$ $SOC\{-R \parallel R \parallel R - R^* \parallel R^* - P-\}$ $SOC\{-R \parallel R - R^* \parallel R^* - P - P-\}$
	$\begin{bmatrix} t^2(\perp \rho_{oo'}) \\ r^3 \end{bmatrix}$	$SOC\{-\widehat{RR} - \widehat{RRR}-\}$	$SOC\{-\widehat{RRR} - \widehat{RRR}-\}$
	$\begin{bmatrix} t^2(\perp R^*) \\ r^3 \end{bmatrix}$	$SOC\{-R^* \parallel R^* - \widehat{RRR}-\}$ $SOC\{-R^* \parallel R^* - \widehat{RRR}-\}$	$SOC\{-R^* \parallel R^* \parallel R^* - \widehat{RRR}-\}$
		$SOC\{-R^* \perp P - \widehat{RRR}-\}$ $SOC\{-R^* \parallel R^* \perp P - \widehat{RR}-\}$	$SOC\{-R^* \parallel R^* \perp P - \widehat{RRR}-\}$
		$SOC\{-P - P - \widehat{RRR}-\}$	$SOC\{-P \perp R^* \perp P - \widehat{RRR}-\}$

Note Base point o′ is on axis of the end pair or is the common interesting point of several R pairs

8.3.2 *Expansion of Structure Types*

Although there are a finite number of structure types containing only R and P pairs (Table 8.1) more structure types containing multi-DOF pairs (such as C, U, S, E etc.) can be obtained through expansion described in Sect. 8.2.3.

8.4 Structure Synthesis of Single-Loop Mechanisms

8.4.1 *Classification of Single-Loop Mechanisms*

As discussed in Sect. 2.6 in Chap. 2, single-loop chain (SLC) can be classified into the following three types:

(1) Non-overconstrained SLC

The number of independent displacement equations of a non-overconstrained SLC is $\xi_L = 6$ and there is DOF $\geq$ 1.

(2) General overconstrained SLC

The number of independent displacement equations of a general overconstrained SLC is $\xi_L = 2 \sim 5$ and there is DOF $\geq$ 1.

A general overconstrained SLC is composed of several geometric constraint types of axes (sub-SOCs, in Table 4.1 in Chap. 4) connected in series. For example, the two mechanisms in Fig. 2.4 in Chap. 2 are general overconstrained SLCs.

(3) Special overconstrained SLC

The number of independent displacement equations of a special overconstrained SLC is $\xi_L = 2 \sim 5$ and there must be DOF = 1.

Existence conditions of a special overconstrained SLC include special function relations among dimensional parameters of the mechanism [18–20]. Bennett mechanism and Bricard mechanism (Fig. 2.7 in Chap. 2) are special overconstrained SLCs.

It should be noted that only the structure synthesis of general overconstrained mechanisms is discussed in the section.

8.4.2 *Structure Synthesis of Single-Loop Mechanisms*

Since a multi-DOF kinematic pair (such as C, S, etc.) can be regarded as being composed of R and P pairs connected in series (Sect. 2.3 in Chap. 2), structure synthesis of SLC containing only R, P and H pairs is discussed in this section [13–17].

(1) Basic requirements of SLC structure synthesis

(a) Achieve the required number of independent displacement equations.
(b) DOF = 1.

(2) Basic idea of the structure synthesis method

As discussed in Sect. 4.6 in Chap. 4, break a certain link of the SLC and obtain the corresponding $SOC_{(SLC)}$. The SLC and the corresponding $SOC_{(SLC)}$ have the same topological structure. The number of independent displacement equations of the SLC (ξ_L) is equal to the dimension of POC set of the $SOC_{(SLC)}$ (Eq. 4.5 in Chap. 4).

On the contrary, if two end links of the corresponding $SOC_{(SLC)}$ are combined into one link. The obtained SLC has the same topological structure as the $SOC_{(SLC)}$. The number of independent displacement equations of the SLC (ξ_L) is equal to the dimension of POC set of the $SOC_{(SLC)}$ ($\dim\{M_{S(L)}\}$).

So, structure synthesis of SLCs can be converted to structure synthesis of serial mechanisms (SOCs).

(3) Main steps of structure synthesis

The main steps are as follows:

Step 1 Determine the dimension of POC set of the corresponding $SOC_{(SLC)}$. According to Eq. (4.5), there is

$$\dim(M_{S(L)}) = \xi_L$$

where, $\dim\{M_{S(L)}\}$—dimension of POC set of the corresponding $SOC_{(SLC)}$, ξ_L—number of independent displacement equations of the SLC.

Step 2 Determine pair combination schemes of the $SOC_{(SLC)}$.
Determine pair combination schemes of the $SOC_{(SLC)}$ according to Eqs. (8.1–8.3), as discussed in step 2 in Sect. 8.2.3.

Step 3 Determine sub-SOCs contained in the $SOC_{(SLC)}$.
Determine sub-SOCs contained in the $SOC_{(SLC)}$, as discussed in step 3 in Sect. 8.2.3.

Step 4 Generate structure types of the $SOC_{(SLC)}$
Connect the sub-SOCs determined in step 3 in series and generate structure types of the $SOC_{(SLC)}$.

Step 5 Check POC set of the $SOC_{(SLC)}$.
For the structure types obtained in step 4, check whether POC set of the $SOC_{(SLC)}$ is equal to the required number of independent displacement equations (ξ_L) based on POC equation for serial mechanism (Eq. 4.4 in Chap. 4).

It should be noted that base point o′ on the end link of $SOC_{(SLC)}$ should be the intersection of axes of several R pairs or should be on axis of R (H) pair (refer to Sect. 4.6 in Chap. 4).

Step 6 Obtain structure types of the SLC
Combine two end links of the $SOC_{(SLC)}$ obtained in step 4 into one link and obtain structure types of the SLC.

Step 7 Determine which pair can be used as the driving pair
Determine which pair can be used as the driving pair of the SLC according to criteria for selection of driving pairs (Sect. 6.3.2 in Chap. 6).

Step 8 Draw the schematic diagram.

Example 8.3 Generate an SLC from the SOC in Fig. 8.3a.

(1) Topological structure of the SOC in Fig. 8.3a:

$$SOC\{-R_1 \parallel R_2 \parallel R_3 \perp R_4 | H_5-\}.$$

It is composed of $sub-SOC\{-R \parallel R \parallel R-\}$ and $sub-SOC\{-R|H-\}$ connected in series and there is $R_3 \perp R_4$.

(2) Point o' on axis of H_5 pair is selected as the base point.

(3) Determine the dimension of POC set of the SOC's end link.

(a) Establish POC equation of this SOC.
Substitute POC sets of $sub-SOC\{-R \parallel R \parallel R-\}$ and $sub-SOC\{-R|H-\}$ (No. 2 and No. 7* in Table 4.1) into POC equation for serial mechanisms (Eq. 4.4), there is

$$M_S = \begin{bmatrix} t^2(\perp R_3) \\ r^1(\parallel R_3) \end{bmatrix} \cup \begin{bmatrix} t^1(\parallel H_5) \\ r^1(\parallel R_4) \end{bmatrix}$$

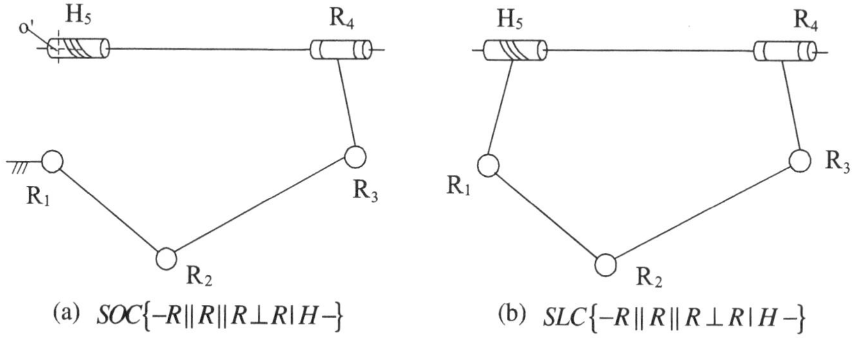

Fig. 8.3 $SLC\{-R \parallel R \parallel R \perp R|H-\}$

(b) "Union" operation among rotational elements
According to operation rules for rotational elements (Eqs. 4.4a–4.4d in Chap. 4), the above equation can be rewritten as

$$M_S = \begin{bmatrix} t^2(\perp R_3) \\ r^1(\| R_3) \end{bmatrix} \cup \begin{bmatrix} t^1(\| H_5) \\ r^1(\| R_4) \end{bmatrix} = \begin{bmatrix} t^2(\perp R_3) \cup t^1(\| H_5) \\ r^2(\| \diamondsuit(R_3, R_4)) \end{bmatrix}$$

(c) "Union" operation among translational elements
According to operation rules for translational elements (Eqs. 4.4e–4.4h) and considering $R_3 \perp H_5$, the above equation can be rewritten as

$$M_S = \begin{bmatrix} t^2(\perp R_3) \\ r^1(\| R_3) \end{bmatrix} \cup \begin{bmatrix} t^1(\| H_5) \\ r^1(\| R_4) \end{bmatrix} = \begin{bmatrix} t^2(\perp R_3) \cup t^1(\| H_5) \\ r^2(\| \diamondsuit(R_3, R_4)) \end{bmatrix}$$
$$= \begin{bmatrix} t^2(\perp R_3) \\ r^2(\| \diamondsuit(R_3, R_4)) \end{bmatrix}$$

So, dimension of the POC set is $\dim\{M_S\} = 4$.

(4) Generate the corresponding SLC
Combine two end links of $SOC\{-R_1 \| R_2 \| R_3 \perp R_4 | H_5-\}$ into one link and obtain $SLC\{-R_1 \| R_2 \| R_3 \perp R_4 | H_5-\}$, as shown in Fig. 8.3(b). There are $\xi_L = 4$ and $DOF = 1$.

(5) Select the driving pair
According to criteria for driving pair selection (Sect. 6.3.2 in Chap. 6), it is easy to determine that any kinematic pair of this SLC can be used as the driving pair.

8.4.3 Structure Types of Single-Loop Mechanisms

Combine two end links of the serial mechanism structure types with $DOF = \dim\{M_S\} + 1$ in Table 8.1 into one link and obtain 15 general overconstrained single-loop kinematic chains with DOF = 1 containing only R and P pairs [16, 17], as shown in Table 8.2.

Table 8.2 General overconstrained single-loop kinematic chains with DOF = 1 [16, 17]

ξ_L	Structure types of single-loop kinematic chains		
2	No. 1 $SLC\{\lozenge(-P-P-P-)\}$ 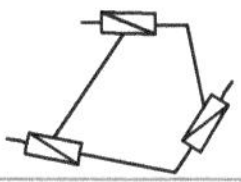		
3	No. 2 $SLC\{-R \parallel R \parallel R \parallel R-\}$ 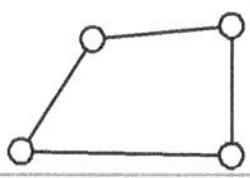	No. 3 $SLC\{-R \parallel R \parallel R \perp P-\}$ 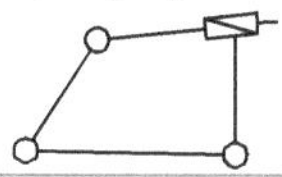	No. 4 $SLC\{-P \perp R(\perp P) \parallel R-\}$ 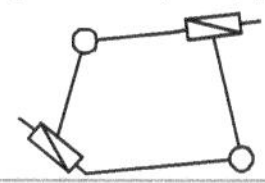
	No. 5 $SLC\{-\overbrace{RRRR}-\}$ 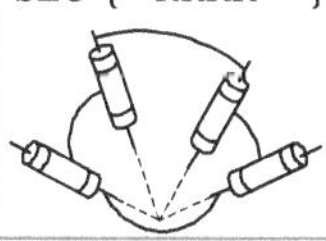	No. 6 $SLC\{-P-P-P-P-\}$ 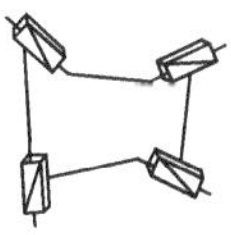	
4	No. 7 $SLC\{-R \parallel R \parallel R-P-P-\}$ 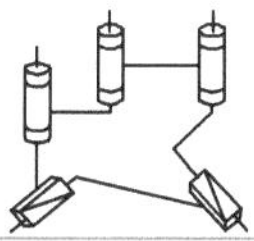	No. 8 $SLC\{-R \parallel R-P-P-P-\}$ 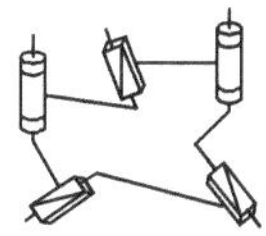	
5	No. 9 $SLC\{-R \parallel R \parallel R-R \parallel R \parallel R-\}$ 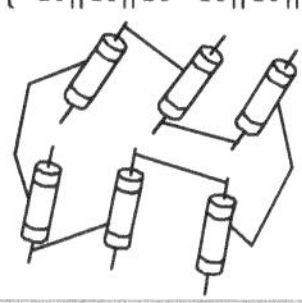	No. 10 $SLC\{-R \parallel R \parallel R-R \parallel R-P-\}$ 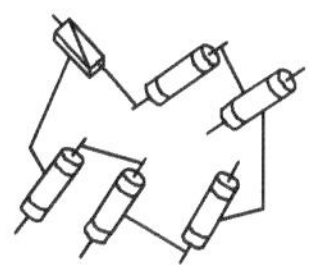	No. 11 $SLC\{-R \parallel R-R \parallel R-P-P-\}$ 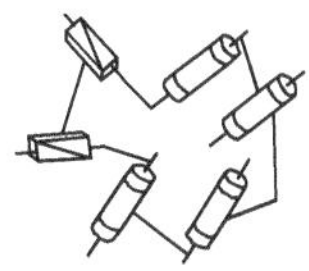
	No. 12 $SLC\{-\overbrace{RRR}-\overbrace{RRR}-\}$ 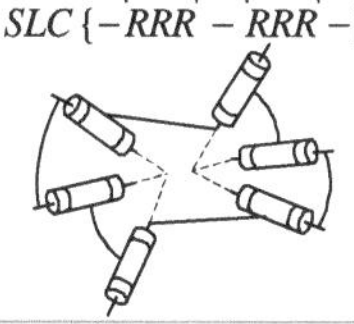		
	No. 13 $SLC\{-R \parallel R \parallel R-\overbrace{RRR}-\}$ 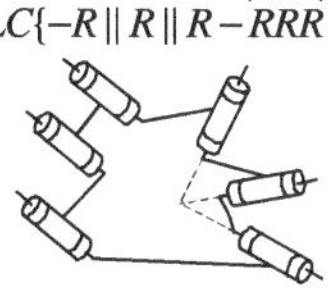	No. 14 $SLC\{-R \parallel R \perp P-\overbrace{RRR}-\}$ 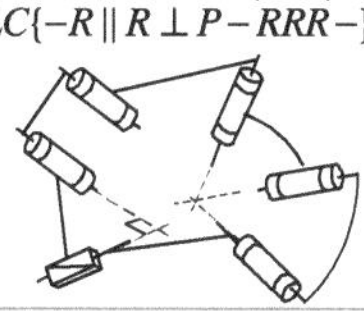	No. 15 $SLC\{-P \perp R \perp P-\overbrace{RRR}-\}$ 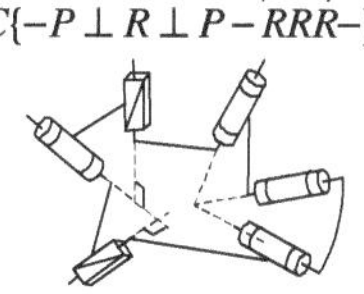

8.5 Summary

(1) A easy-to-use method for structure synthesis of serial mechanisms is proposed based on POC equation for serial mechanisms (Eq. 4.4 in Chap. 4) and sub-SOCs of geometric constraint types of axes (Table 4.1 in Chap. 4).

(2) 28 structure types of serial mechanisms containing only R and P pairs with DOF = dim$\{M_S\}$ = 2–5 and 15 structure types of serial mechanisms containing only R and P pairs with DOF = dim$\{M_S\}$ + 1 are obtained (Table 8.1). The method for expanding these structure types with C, U and S pairs is given (Sect. 8.2.3).

(3) Based on the one-to-one correspondence between single-loop chain (SLC) and corresponding $SOC_{(SLC)}$, a simple method for structure synthesis of general overconstrained SLC is proposed. 15 types of one DOF general overconstrained SLCs containing only R and P pairs are obtained (Table 8.2).

(4) The method for structure synthesis of serial mechanisms is a geometrical method which is independent of motion position of mechanisms and the fixed coordinate system (i.e. it is not necessary to establish the fixed coordinate system).

(5) Contents in this chapter provide a theoretical basis for structure synthesis of simple and complex branches of PMs in chap. 9.

References

1. Hunt KH (1987) Kinematic geometry of mechanisms. Clarendon Press, Oxford
2. Huang Z, Li QC (2002) General methodology for type synthesis of symmetrical lower-mobility parallel manipulators and several novel manipulators. Int J Robot Res 21:131–145
3. Kong X, Gosselin C (2007) Type synthesis of parallel mechanisms. Springer, Heidelberg
4. Herve JM (1999) The Lie group of rigid body displacements, a fundamental tool for mechanism design. Mech Mach Theor 34:719–730
5. Meng J, Liu GF, Li ZX (2007) A geometric theory for analysis and synthesis of sub-6 DOF parallel manipulators. IEEE Trans Rob 23:625–649
6. Meng X, Gao F, Wu S, Ge J (2014) Type synthesis of parallel robotic mechanisms: framework and brief review. Mech Mach Theor 78:177–186
7. Gogu G (2007) Structural synthesis of parallel robots: part 1: methodology. Springer, Dordrecht
8. Yang T-L, Jin Q et al (2001) A general method for type synthesis of rank-deficient parallel robot mechanisms based on SOC unit. Mach Sci Technol 20(3):321–325
9. Yang T-L, Jin Q, Liu A-X, Shen H-P, Luo Y-F (2002) Structural synthesis and classification of the 3-DOF translation parallel robot mechanisms based on the unites of single-open chain, Chinese. J Mech Eng 38(8):31–36. doi:10.3901/JME.2002.08.031
10. Jin Q, Yang T-L (2004) Theory for topology synthesis of parallel manipulators and its application to three-dimension-translation parallel manipulators. ASME J Mech Des 126:625–639

11. Yang T-L (2004) Theory of topological structure for robot mechanisms. China machine press, Beijing
12. Yang T-L, Liu A-X, Luo Y-F et al (2009) Position and orientation characteristic equation for topological design of robot mechanisms. ASME J Mech Design 131:021001
13. Yang T-L, Sun D-J (2008) Rank and mobility of single loop kinematic chains. In: Proceedings of the ASME 32-th mechanisms and robots conference, DETC2008-49076
14. Yang T-L, Sun D-J (2012) A general DOF formula for parallel mechanisms and multi-loop spatial mechanisms. ASME J Mech Robot. 4:011001
15. Yang T-L, Liu A-X, Shen H-P et al (2013) On the correctness and strictness of the position and orientation characteristic equation for topological design of robot mechanisms. ASME J Mech Robot 5:021009
16. Yang T-L (1986) Kinematic structural analysis and synthesis of over-constrained spatial single loop chains. In: Proceedings of the 19th Biennial mechanisms conference on Columbus ASME paper 86-DET-189
17. Yang T-L, Liu A-X, Luo Y-F et al (2012) Theory and application of robot mechanism topology. China Science Press, Beijing
18. Baker JE (1979) The Bennett, Goldberg and Myard linkages—in perspective. Mecha Mach Theor 14:239–253
19. Baker JE (1980) An analysis of the Bricard linkages. Mech Mach Theor 15:267–286
20. Jin Q, Yang T-L (2002) Over-constraint analysis on spatial 6-link loops. Mech Mach Theor 37(3):267–278

Chapter 9
General Method for Topology Design of Parallel Mechanisms

9.1 Introduction

9.1.1 Basic Mode of Mechanism Topology Design

Generally, the mechanism topology design follows a 3-stage mode:

⟪Task space⟫	→	⟪Solution space⟫	→	⟪Solution optimization⟫
(Basic functions)		(Set of structure types)		(Structure type optimization)

(1) Task space: to determine the design objectives based on the design specification, including functional requirements and performance requirements.

(2) Solution space: to obtain many structure types which meet the basic functional requirements using the structure synthesis method and thus to provide a relatively larger solution space for structure type optimization.

(3) Solution optimization: to carry out performance analysis, comparison and classification to the structure types in the solution space based on functional requirements, performance requirements and other specific requirements and to provide some useful information for structure type optimization.

9.1.2 Basic Requirements for Topology Design

Basic requirements for topology design include:

(1) POC set of the movable platform and dimension of this POC set.

(2) DOF of the parallel mechanism.

(3) Other requirements, such as

T.-L. Yang et al., *Topology Design of Robot Mechanisms*,
https://doi.org/10.1007/978-981-10-5532-4_9

(a) All driving pairs are located on the same platform or are as close to the same platform as possible.
(b) Each branch contains one P pair at most and this P pair is the driving pair.
(c) Complexity of the kinematic (dynamic) analysis.
(d) Decoupling between motion inputs and outputs.
(e) Symmetry of the topological structure.

9.1.3 Main Method for Topology Design of Parallel Mechanisms

In the last decade, topology design of parallel robot mechanism became the research focus of many researchers. There formed the following four categories of topology design methods:

(1) Method based on screw theory [1–10]

Main features of this method as follows:

(a) Six components of plucker coordinates are used to describe relative motion characteristics between two links of a mechanism. They are dependent on motion position of the mechanism and the fixed coordinate system.

(b) Linear operation of the screw system is relatively simple.

(c) Complex 'intersection' operation among motion screw systems is converted to simple 'union' operation among constraint screw systems by using reciprocal product of screws [3, 4].

(d) For the obtained parallel mechanisms, it is necessary to carry out full-cycle DOF verification [3–10].

(2) Method based on subgroup/submanifold [11–23]

Main features of this method as follows:

(a) 12 types of displacement subgroups are used to describe relative motion characteristics between two links of a mechanism fulfilling the algebra structures of Lie group [11, 12]. They are independent of motion position of the mechanism.

(b) 'Union' and 'intersection' operation rules are provided for the 12 types of displacement subgroups using synthetic arguments or tables of compositions of subgroups [11, 14].

(c) The full-cycle DOF mechanisms are obtained.

(d) Submanifold mechanism which does not have a displacement subgroup structure is not cover by this method [11]. So, some researchers now use differentiable manifold to develop the subgroup/submanifold based method, which depends on motion position of the mechanism. It is necessary to carry out full cycle DOF verification for obtained mechanisms [21].

(3) Method based on theory of linear transformations and evolutionary morphology [24–29]

Main features of this method as follows:

(a) Velocity space is used to describe relative motion characteristics between two links of a mechanism. It is dependent on motion position of the mechanism and the fixed coordinate system.

(b) The design objectives include DOF of branch, DOF of PM, connectivity of branch between the moving platform and the fixed platform, connectivity of PM between the moving platform and the fixed platform, number of overconstraints and redundancy of the mechanism.

(c) Morphological operations (the combination, the mutation, the migration and the selection) and evolutionary rules are used to determine type, number and order of kinematic pairs and allocation of pair axes.

(d) For the obtained parallel mechanisms, it is necessary to carry out ful cycle DOF verification.

(4) Method based on POC equations [30–50]

Main features of this method as follows:

(a) Six elements of POC set are used to describe relative motion characteristics between two links of a mechanism. They are independent of motion position of the mechanism (excluding singular position) and the fixed coordinate system.

(b) The topology design process is divided into two stages: the structure synthesis is conducted and many structure types are obtained during the first stage, and performance analysis and classification are carried out for those obtained structure types during the second stage.

(c) Theoretical bases of the topology design are the POC equation for serial mechanisms, the POC equation for parallel mechanisms and the general DOF formula. These equations (formula) are independent of motion position of a mechanism and the fixed coordinate system.

(d) Since this method is independent of motion position, the full-cycle DOF mechanisms are obtained. Since the method is independent of the fixed coordinate system, the geometry conditions of mechanism existence have generality.

9.2 General Procedure for Topology Design of Parallel Mechanisms

General procedure for topology design of parallel mechanisms is shown in Fig. 9.1. The whole design process includes two stages. The first stage involves structure synthesis. Many different structure types can be obtained in this stage. The second stage includes performance analysis and classification of PM structure types obtained in the first stage. Result of the second stage can be used for optimum selection of structure types. Each step of this procedure is very easy to follow since there are explicit formulas or criteria for it.

The following four key steps will be discussed in detail in Sects. 9.3–9.6:

(1) Structure syntheses of SOC branches and HSOC branches.

(2) Number of branches and branch combination schemes.

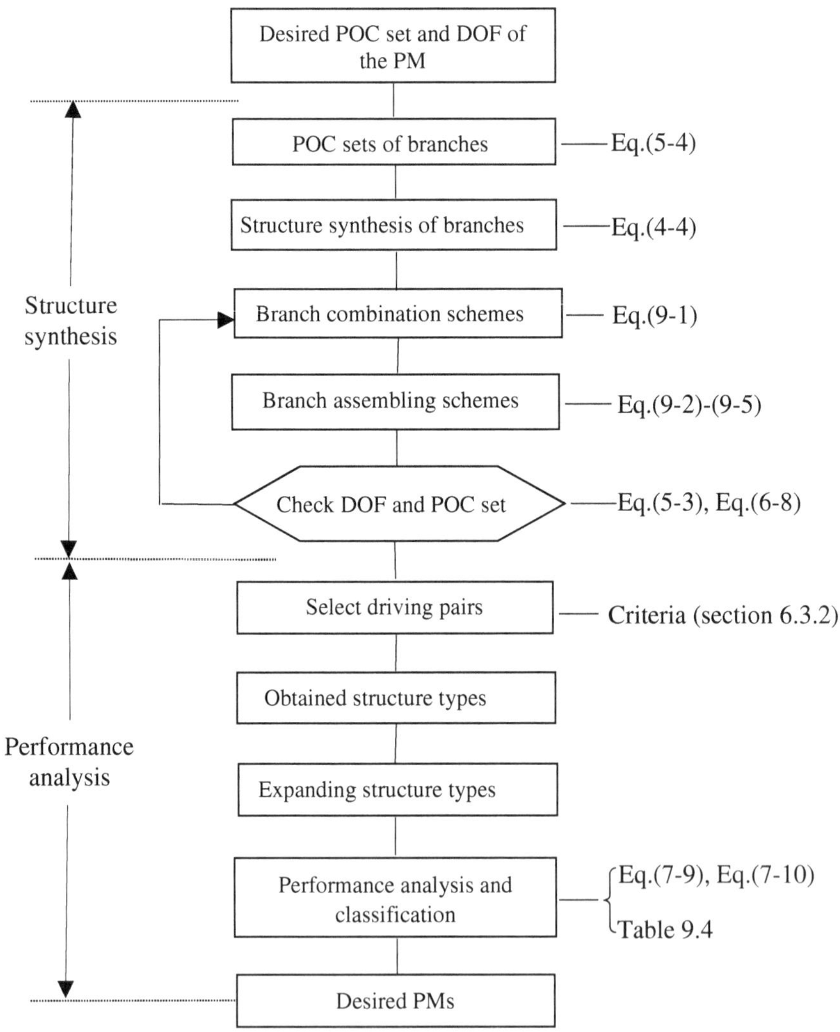

Fig. 9.1 General procedure for topology design of parallel mechanisms

(3) Geometrical conditions and schemes for assembling branches between two platforms.

(4) Performance analysis and classification of PM structure types.

9.3 Structure Synthesis of Branches

9.3.1 Basic Requirements for Structure Synthesis of Branches

(1) POC set of the branch.
When POC set of the PM (M_{Pa}) is known, POC set of the branch (M_{b_j}) can be determined according to Eq. (5.4) in Chap. 5.

(2) DOF of the branch is equal to dimension of the branch's POC set. It means that the branch has no redundant DOF.

(3) All driving pairs are located on the same platform or are as close to the same platform as possible.

(4) P pair in the branch must be used as the driving pair.

Example 9.1 If POC set of the PM is $M_{Pa} = \begin{bmatrix} t^3 \\ r^0 \end{bmatrix}$, determine POC set of the branch.

According to Eq. (5.4) in Chap. 5 (i.e., $M_{b_j} \supseteq M_{Pa}$), POC set of the branch can be selected as

$$M_{b_j} = \begin{bmatrix} t^3 \\ r^0 \end{bmatrix}, \begin{bmatrix} t^3 \\ r^1 \end{bmatrix}, \begin{bmatrix} t^3 \\ r^2 \end{bmatrix}, \begin{bmatrix} t^3 \\ r^3 \end{bmatrix}.$$

Correspondingly, the branch can be noted as (3T-0R), (3T-1R), (3T-2R) or (3T-3R) branch.

9.3.2 Structure Synthesis of SOC Branches

(1) Structure synthesis of SOC branches

If POC set of the SOC branch (i.e. simple branch) is known, many structure types of the SOC branch containing only R and P pairs can be obtained using the structure synthesis method based on POC equation for serial mechanisms described in Chap. 8.

Since structure types of SOC branch containing only R and P pairs are listed in Table 8.1 in Chap. 8, the SOC branch with corresponding POC set can be selected directly from Table 8.1. For the above four possible POC sets of the branch in Example 9.1, 11 structure types of the SOC branch with DOF = $\dim\{M_S\}$ are selected from Table 8.1, as shown in Table 9.1.

Table 9.1 Structure types of the SOC branch

M_{b_j}	SOC branch	
(3T-0R)	1	$SOC\{-P-P-P-\}$
(3T-1R)	2	$SOC\{-R \parallel R \parallel R-P-\}$
	3	$SOC\{-R \parallel R-P-P-\}$
	4	$SOC\{-R-P-P-P-\}$
(3T-2R)	5	$SOC\{-R \parallel R \parallel R-R \parallel R-\}$
	6	$SOC\{-R \parallel R \parallel R-R-P-\}$
	7	$SOC\{-R \parallel R-R \parallel R-P-\}$
	8	$SOC\{-R \parallel R-R-P-P-\}$
	9	$SOC\{-R-R-P-P-P-\}$
(3T-3R)	10	$SOC\{-S-S-R-\}$
	11	$SOC\{-S-S-P-\}$

(2) Expansion of the structure types of the SOC branch
For SOC branches in Table 9.1, new SOC branches can be generated using the following methods:

(a) Change connection order of kinematic pairs
For example, four new SOC branches can be obtained through changing connection order of kinematic pairs to the No. 3 SOC branch in Table 9.1.

① $SOC\{-R \parallel R-P-P-\}$.

② $SOC\{-R(-P) \parallel R-P-\}$, axes of the two R pairs are parallel.

③ $SOC\{-R(-P-P) \parallel R-\}$, axes of the two R pairs are parallel.

④ $SOC\{-P-R \parallel R-P-\}$.

(b) Combine several R pairs and P pairs into a C (or U) pair.
For example, a new branch $SOC\{-R \parallel R \parallel C-\}$ can be obtained through combining the R pair and the P pair into a C pair. A new branch $SOC\{-R \parallel R \parallel \widehat{R \perp R} \parallel R-\}$ can be obtained through combining the third R pair and the fourth R pair into a U pair. $\widehat{R \perp R}$ is the symbolic representation of U pair (refer to Chap. 2).

(c) Replace several R pairs with an S pair.
For example, $SOC\{-\widehat{RRR}-\}$ in which axes of the three R pairs intersect at a common point can be replaced by an S pair.

Although there are only a finite number of SOC branches containing only R and P pairs (Table 9.1). Much more SOC branches containing multi-DOF kinematic pairs (C, U, S, etc.) can be obtained through expansion. Thus the solution space for structure synthesis of SOC branch can be enlarged.

9.3.3 *Structure Synthesis of HSOC Branches*

Structure synthesis of complex branches (i.e. Hybrid SOCs, in short, HSOCs) based on topological equivalence principle is introduced in this section. The HSOC branch can be used to make it possible to arrange all driving pairs on the same platform and to improve kinematic, dynamic and control performance.

A. **Topological equivalence principle**

(1) Topologically equivalent kinematic chains

If two kinematic chains have the same POC set, they are topologically equivalent kinematic chains.

For example, the SOC branch in Fig. 9.2a and the HSOC branch in Fig. 9.2b are topologically equivalent since they have the same POC set:

$$M_S = \begin{bmatrix} t^3 \\ r^1 \end{bmatrix}$$

The sub-$SOC\{-P-P-P-\}$ in Fig. 9.2a and sub-PM (Delta mechanism) in Fig. 9.2b are topologically equivalent since they have the same POC set:

$$M_{sub-S} = M_{sub-PM} = \begin{bmatrix} t^3 \\ r^0 \end{bmatrix}$$

(2) Topological equivalence principle

Generally, if the sub-SOC of an SOC branch is replaced with a topologically equivalent sub-PM, an HSOC branch can be obtained. Since the sub-SOC and the sub-PM have the same POC set, the HSOC branch and the original SOC branch also have the same POC set. This is called topological equivalence principle.

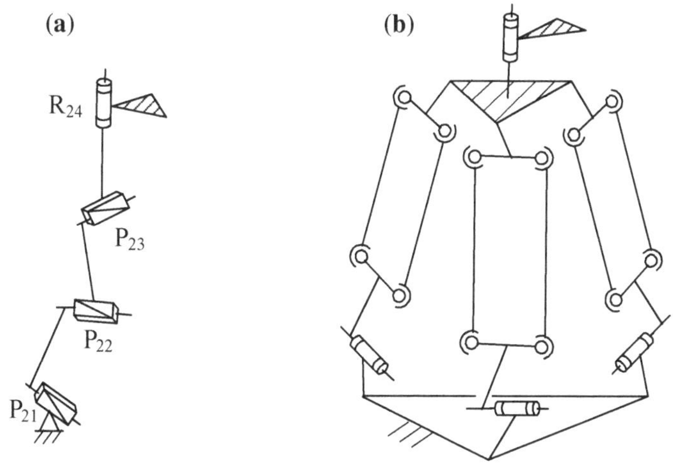

Fig. 9.2 Two topologically equivalent branches

Based on the topological equivalence principle, for any SOC branch, the corresponding HSOC with the same POC set can be obtained by replacing its sub-SOC with topologically equivalent sub-PM.

Since most HSOCs contains only two-branch sub-PMs, seven kinds of two-branch sub-PMs and their topologically equivalent sub-SOCs are listed in Table 9.2.

It should be noted that any parallel mechanism can be used as a sub-PM to replace a sub-SOC with the same POC set.

B. **Structure synthesis of HSOC branches**

Main steps for structure synthesis as follows:

Step 1 Determine POC set and DOF of the HSOC branch.

Step 2 Select the SOC branch which has the same POC set and DOF as the HSOC branch.

Step 3 Select the sub-SOC of the SOC branch and replace it with the topologically equivalent sub-PM. The HSOC branch can be obtained

Example 9.2 Generate the HSOC branch containing no P pairs from No. 3 SOC in Table 9.1 (i.e. $SOC\{-R \parallel R - P - P-\}$).

The following are three such HSOC branches:

(1) Replace sub-$SOC\{-R - P - P-\}$ of $SOC\{-R \parallel R - P - P-\}$ with topologically equivalent sub-$PM\{-R^{(4S)} \parallel P^{(4S)} \perp P^{(4S)}-\}$ (No. 3 in Table 9.2) and $HSOC\{-R \parallel (R^{(4S)} \parallel P^{(4S)} \perp P^{(4S)})-\}$ can be obtained (Fig. 9.3b). Since No. 2 SOC in Table 9.1 (Fig. 9.3a) and the No. 3 SOC are topologically equivalent, the branch in Fig. 9.3a and the branch in Fig. 9.3b are topologically equivalent.

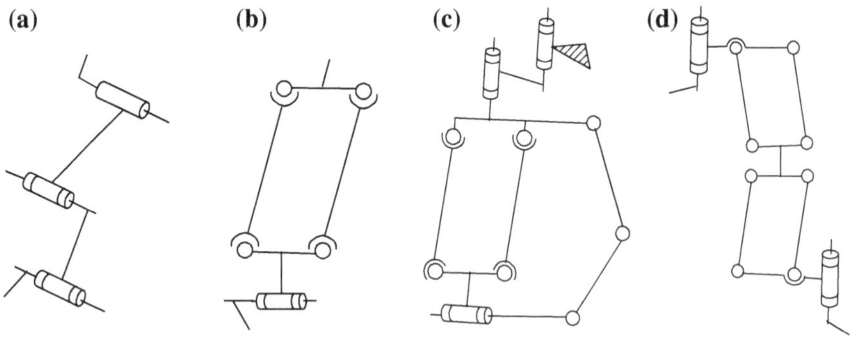

Fig. 9.3 Topologically equivalent branches of the No. 3 SOC in Table 9.1

Table 9.2 Sub-PMs and their topologically equivalent sub-SOCs [37–44]

No.	Sub-PMs	POC set	sub-SOC
1	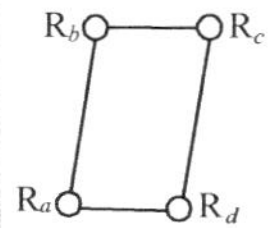 $HSOC\{-P^{(4R)}-\}$	$\begin{bmatrix} t^1(\Vert^{\perp}(ad)) \\ r^0 \end{bmatrix}$	 $SOC\{-P-\}$
2	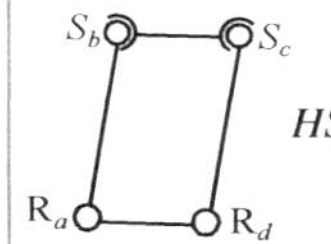 $HSOC\{-R^{(2S2R)} \Vert^{\perp} P^{(2S2R)}-\}$	$\begin{bmatrix} t^1(\Vert^{\perp}(ad)) \\ r^1(\Vert(bc)) \end{bmatrix}$	 $SOC\{-R \Vert^{\perp} P-\}$
3	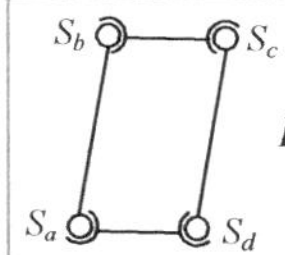 $HSOC\{-R^{(4S)} \Vert P^{(4S)} \perp P^{(4S)}-\}$	$\begin{bmatrix} t^1(\Vert(ad)) \quad t^1(\perp(ad)) \\ r^1(\Vert(ad)) \end{bmatrix}$	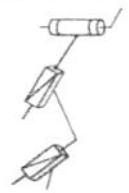 $SOC\{-R \Vert P \perp P-\}$
4	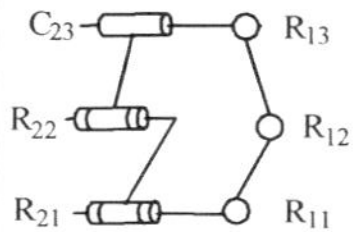 $HSOC\{-\Diamond(P^{(2R1C-3R)}, P^{(2R1C-3R)})-\}$	$\begin{bmatrix} t^2(\perp R_{11}) \\ r^0 \end{bmatrix}$	 $SOC\{-\Diamond(P,P)-\}$
5	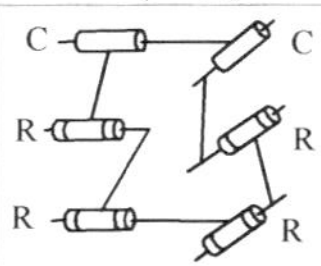 $HSOC\{-P^{(2-RRC)}-P^{(2-RRC)}-P^{(2-RRC)}-\}$	$\begin{bmatrix} t^3 \\ r^0 \end{bmatrix}$	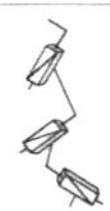 $SOC\{-P-P-P-\}$

(continued)

Table 9.2 (continued)

No.	Sub-PMs	POC set	sub-SOC
6	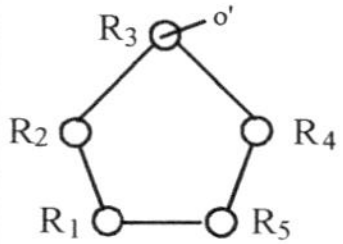 $HSOC\{-\diamondsuit(P^{(5R)}, P^{(5R)})-\}$	$\begin{bmatrix} t^2(\perp R) \\ r^0 \end{bmatrix}$	 $SOC\{-\diamondsuit(P, P)-\}$
7	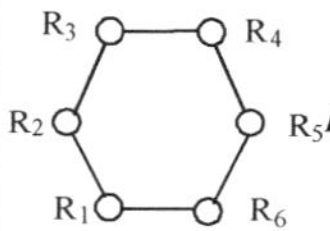 $HSOC\{-R^{(6R)}\perp\diamondsuit(P^{(6R)}, P^{(6R)})-\}$	$\begin{bmatrix} t^2(\perp R) \\ r^1(\parallel R) \end{bmatrix}$	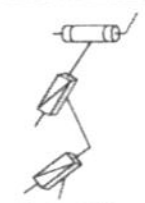 $SOC\{-R\perp\diamondsuit(P, P)-\}$

Note The four S pairs of No. 3 sub-PM should form a parallelogram all the time

(2) Replace sub-$SOC\{-P-P-\}$ of $SOC\{-R \parallel R-P-P-\}$ with topologically equivalent sub-sub-$PM\{-\diamondsuit(P^{(2R1C-3R)}, P^{(2R1C-3R)})-\}$ (No. 4 in Table 9.2) and $HSOC\{-R \parallel R\bot(\diamondsuit(P^{(2R1C-3R)}, P^{(2R1C-3R)})-\}$ can be obtained. Then replace sub-$SOC\{-R \parallel R \parallel C-\}$ (No. 4 in Table 9.2, Fig. 9.3a) with topologically equivalent $HSOC\{-R \parallel (R^{(4S)} \parallel P^{(4S)}\bot P^{(4S)})-\}$ (Fig. 9.3b). The obtained HSOC branch is shown in Fig. 9.3c.

(3) Replace two sub-$SOC\{-P-\}$ of $SOC\{-R \parallel R-P-P-\}$ with topologically equivalent sub- sub-$PM\{-P^{(4P)}-\}$ (No. 1 in Table 9.2) and change connection order of kinematic pairs, $HSOC\{-R(\parallel \diamondsuit(P^{(4R)}, P^{(4R)}) \parallel R-\}$ can be obtained. $\diamondsuit(P^{(4R)}, P^{(4R)})$ means that the two parallelograms are coplanar and axes of two R pairs are parallel, as shown in Fig. 9.3d.

According to this example, for the same SOC branch, many different HSOCs can be obtained. Use of topologically equivalent sub-PMs may further enlarge the solution space of HSOC structure synthesis.

9.4 Branch Combination Schemes

Many structure types of SOC branch and HSOC branch can be obtained during structure synthesis of branches. In order for the PM to achieve the prescribed POC set, number of branches and branch combination schemes should be determined.

9.4.1 Number of Branches

Number of branches of the PM depends on the following factors:

(1) Topological structure and POC set of the branch (M_b) are known. In order for the PM to achieve the prescribed POC set (M_{Pa}) through "intersection" of each branch's POC set, the minimum number of branches $n_{b-\min}$ is

$$n_{b-\min} = \dim\{M_b\} - \dim\{M_{Pa}\} + 1. \tag{9.1}$$

Equation (9.1) means that in order for the PM to achieve the prescribed POC set (M_{Pa}), at least $n_{b-\min}$ branches are needed. "Intersection" of the POC sets of these $n_{b-\min}$ branches can eliminate those elements which are not included in POC set of the PM.

For example, the minimum number of branches for the No. 2 SOC in Table 9.1 ($SOC\{-R \parallel R \parallel R - P-\}$) to achieve three translational outputs is

$$n_{b-\min} = \dim\{M_b\} - \dim\{M_{Pa}\} + 1 = 4 - 3 + 1 = 2.$$

The minimum number of branches for the No. 5 SOC in Table 9.1 ($SOC\{-R \parallel R \parallel R - R \parallel R-\}$) to achieve three translational outputs is

$$n_{b-\min} = \dim\{M_b\} - \dim\{M_{Pa}\} + 1 = 5 - 3 + 1 = 3$$

(2) All driving pairs are located on the same platform.

   (a) If each branch has only one driving pair, the number of branches is equal to DOF of the PM.

   (b) If the PM contains HSOC branches, the number of frame connecting pairs in all branches should be equal to DOF of the PM.

(3) In order that all driving pairs can be located on the same platform, the number of branches may sometimes be greater than DOF of the PM.

For example, if the branch contains only one S pair, three other branches are needed to make it possible to allocate all driving pairs on the same platform, as shown in Fig. 9.4.

(4) The number of branches is equal to the prescribed number of pairs on the movable platform.

For example, it is sometimes required that the movable platform have only two kinematic pairs [23].

(5) If any redundant branches are necessary, the number of branches should be increased.

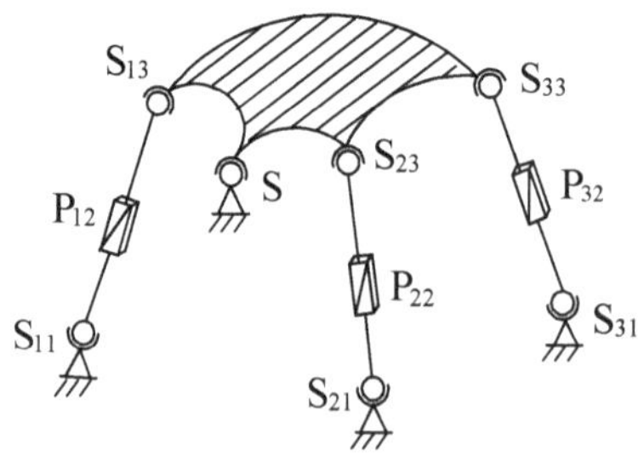

Fig. 9.4 A four-branch spherical mechanism

9.4.2 Branch Combination Schemes

Determination of branch combination schemes should consider several different factors: number of branches, driving pairs allocation, number of driving pairs in HSOC branch, motion decoupling property, number of pairs on movable platform, complexity of branch structure, symmetry of branches, etc. The number of branches is the first factor to be considered (Sect. 9.4.1). The following principled should be followed:

(1) All branches have the same structure type, i.e. the mechanism has topological structure symmetry.

(2) A part of the branches have the same structure type and have already achieved the PM's POC set. Thus, other branches may use relative simple structure types, such as $SOC\{-S-S-R-\}$.

(3) HSOC branch is used in order to for the PM to achieve input-output motion decoupling (refer to chap. 7).

Example 9.3 Determine branch combination schemes of the PM whose POC set contains only three translations.

In order to make the determination process simpler and clearer, only two structure types $SOC\{-R \parallel R \parallel C-\}$ (No. 2 SOC in Table 9.1) and $SOC\{-R \parallel R \parallel \widehat{R \perp R} \parallel R-\}$ (No. 5 SOC in Table 9.1) are considered.

(1) Three branches have the same structure type. There are the following two branch combination schemes:

① $3-SOC\{-R \parallel R \parallel C-\}$ (No. 1 in Table 9.3).

② $3-SOC\{-R \parallel R \parallel \widehat{R \perp R} \parallel R-\}$ (No. 2 in Table 9.3).

(2) Two branches have the same structure type and have already achieved the PM's POC set. Another branch is $SOC\{-S-S-R-\}$. There is one branch combination scheme:

① $2-SOC\{-R \parallel R \parallel C-\} \oplus 1-SOC\{-S-S-R-\}$ (No. 6 in Table 9.3).

(3) For the $SOC\{-R \parallel R \parallel C-\}$, several HSOC branches can be generated using the method described in Sect. 9.3.3, such as $HSOC\{-R \parallel R \parallel (P^{(4R)}) \parallel R-\}$ (Fig. 5.2b in Chap. 5), $HSOC\{-R \parallel (R^{(4S)} \parallel P^{(4S)} \perp P^{(4S)})-\}$ (Fig. 9.3b), etc. These HSOC branches may contribute more branch combination schemes, such as the No. 3, 4, 5, 7, 8 and 9 in Table 9.3.

Table 9.3 Branch combination schemes

All branches have the same structure	
1	$3-SOC\{-R \parallel R \parallel C-\}$
2	$3-SOC\{-R \parallel R \parallel \overbrace{R\perp R} \parallel R-\}$
3	$3-HSOC\{-R \parallel R \parallel (P^{(4R)}) \parallel R-\}$
4	$3-HSOC\{-R \parallel R \parallel (R^{(2S2R)} \parallel P^{(2S2R)})-\}$
5	$3-HSOC\{-R \parallel (R^{(4S)} \parallel P^{(4S)}\perp P^{(4S)})-\}$
…	
Some branches have the same structure	
6	$2-SOC\{-R \parallel R \parallel C-\}\oplus 1-SOC\{-S-S-R-\}$
7	$2-HSOC\{-R \parallel R \parallel (P^{(4R)}) \parallel R-\}\oplus 1-SOC\{-S-S-R-\}$
8	$2-HSOC\{-R \parallel R \parallel (R^{(2S2R)} \parallel P^{(2S2R)})-\}\oplus 1-SOC\{-S-S-R-\}$
9	$2-HSOC\{-R \parallel (R^{(4S)} \parallel P^{(4S)}\perp P^{(4S)})-\}\oplus 1-SOC\{-S-S-R-\}$
…	

It should be noted that Table 9.3 lists only a part of branch combination schemes in Example 9.3. More branch structure types will contribute more branch combination schemes.

9.5 Branch Assembling Schemes

9.5.1 Determination of Geometrical Conditions for Assembling Branch

Since POC set of each branch contains more (or at least same) elements than POC set of the PM, branches should be so assembled that those elements which are not included in the PM's POC set can be eliminated.

According to operation rules of POC equation for parallel mechanisms (Eq. 5.3 in Chap. 5) [34–43], the basic formulas used in determination of geometrical conditions for branch assembling shall be discussed in two different cases:

(1) **Case 1**

If POC set of the sub-PM formed by the first k branches $(M_{Pa(1-k)})$ still contains any element which is not included in the PM's POC set (M_{Pa}), "intersection" of the POC set of the (k + 1)th branch $(M_{b_{(k+1)}})$ and POC set of the sub-PM $(M_{Pa(1-k)})$ should be able to eliminate this element.

A. **"Intersection" operation which can eliminate rotational element**

Among the operation rules of POC equation for parallel mechanisms (Eq. 5.3 in Chap. 5), the following three operation rules can eliminate rotational elements.

$$\left[r^1(\| R_i)\right]_{bi} \cap \left[r^1(\| R_j)\right]_{bj} = \left[r^0\right]_{Pa}, \quad if\ R_i \nparallel R_j \tag{9.2a}$$

$$\left[r^1(\| R_i)\right]_{bi} \cap \left[r^2(\| \diamondsuit(R_{j1}, R_{j2}))\right]_{bj} = \left[r^0\right]_{Pa}, \quad if\ R_i \nparallel (\diamondsuit(R_{j1}, R_{j2})) \tag{9.2b}$$

$$\left[r^2(\| \diamondsuit(R_1, R_2))\right]_{bi} \cap \left[r^2(\| \diamondsuit(R_{j1}, R_{j2}))\right]_{bj} = \left[r^1(\| (\diamondsuit(R_{i1}, R_{i2}) \cap (\diamondsuit(R_{j1}, R_{j2}))))\right]_{Pa}, \quad if\ (\diamondsuit(R_{i1}, R_{i2})) \nparallel (\diamondsuit(R_{j1}, R_{j2})) \tag{9.2c}$$

Obviously, if the geometrical condition on right side of each equation (Eqs. 9.2a–9.2c) is met, one rotational element can be eliminated.

B. **"Intersection" operation which can eliminate translational element**

Among the operation rules of POC equation for parallel mechanisms (Eq. 5.3 in Chap. 5), the following three operation rules can eliminate translational elements.

$$\left[t^1(\| P_i^*)\right]_{bi} \cap \left[t^1(\| P_j^*)\right]_{bj} = \left[t^0\right]_{Pa}, \ if\ P_i^* \nparallel P_j^* \tag{9.3a}$$

$$\left[t^1(\| P_i^*)\right]_{bi} \cap \left[t^2(\| \diamondsuit(P_{j1}^*, P_{j2}^*)\right]_{bj} = \left[t^0\right]_{Pa}, \ if\ P_i^* \nparallel \diamondsuit(P_{j1}^*, P_{j2}^*) \tag{9.3b}$$

$$\left[t^2(\| \diamondsuit(P_1^*, P_2^*))\right]_{bi} \cap \left[t^2(\| \diamondsuit(P_{j1}^*, P_{j2}^*))\right]_{bj} = \left[t^1(\| (\diamondsuit(P_{i1}^*, P_{i2}^*) \cap (\diamondsuit(P_{j1}^*, P_{j2}^*)))\right]_{Pa}, \ if\ (\diamondsuit(P_{i1}^*, P_{i2}^*)) \nparallel (\diamondsuit(P_{j1}^*, P_{j2}^*)) \tag{9.3c}$$

where, P^*—translation of a P pair or derivative translation of a R(H) pair.

Obviously, if the geometrical condition on right side of each equation (Eqs. 9.3a–9.3c) is met, one rotational element can be eliminated.

(2) **Case 2**

If POC set of the sub-PM formed by the first k branches $(M_{Pa(1-k)})$ is the same as the PM's POC set (M_{Pa}), "intersection" of the POC set of the (k + 1)th branch $(M_{b_{(k+1)}})$ and POC set of the sub-PM $(M_{Pa(1-k)})$ should not eliminate any element.

C. **"Intersection" operation which cannot eliminate rotational element**

Among the operation rules of POC equation for parallel mechanisms (Eq. 5.3 in Chap. 5), the following six operation rules cannot eliminate rotational elements.

$$\left[r^1(\| R_i)\right]_{bi} \cap \left[r^1(\| R_j)\right]_{bj} = \left[r^1(\| R_i)\right]_{Pa},\ if\ R_i \| R_j \tag{9.4a}$$

$$\left[r^1(\| R_i)\right]_{bi} \cap \left[r^2(\| \diamondsuit(R_{j1}, R_{j2}))\right] = \left[r^1(\| R_i)\right]_{Pa}, if\ R_i \| (\diamondsuit(R_{j1}, R_{j2})) \tag{9.4b}$$

$$\begin{aligned}\left[r^2(\| \diamondsuit(R_1, R_2))\right]_{bi} \cap \left[r^2(\| \diamondsuit(R_{j1}, R_{j2}))\right] &= \left[r^2(\| \diamondsuit R_{i1}, R_{i2})\right]_{Pa}, \\ &if (\diamondsuit(R_{i1}, R_{i2})) \| (\diamondsuit(R_{j1}, R_{j2}))\end{aligned} \tag{9.4c}$$

$$\left[r^1(\| R_i)\right]_{bi} \cap \left[r^3\right]_{bj} = \left[r^1(\| R_i)\right]_{Pa} \tag{9.4d}$$

$$\left[r^2(\| \diamondsuit(R_{i1}, R_{i2}))\right]_{bi} \cap \left[r^3\right]_{bj} = \left[r^2(\| \diamondsuit(R_{i1}, R_{i2}))\right]_{Pa} \tag{9.4e}$$

$$\left[r^3\right]_{bi} \cap \left[r^3\right]_{bj} = \left[r^3\right]_{Pa} \tag{9.4f}$$

Obviously, if the geometrical condition on right side of each equation (Eqs. 9.4a–9.4f) is met, no rotational element will be eliminated.

D. **"Intersection" operation which cannot eliminate translational element**

Among the operation rules of POC equation for parallel mechanisms (Eq. 5.3 in Chap. 5), the following six operation rules cannot eliminate translational elements.

$$\left[t^1(\| P_i^*)\right]_{bi} \cap \left[t^1(\| P_j^*)\right]_{bj} = \left[t^1(\| P_i^*)\right]_{Pa}, if\ P_i^* \| P_j^* \tag{9.5a}$$

$$\left[t^1(\| P_i^*)\right]_{bi} \cap \left[t^2(\| \diamondsuit(P_{j1}^*, P_{j2}^*))\right] = \left[t^1(\| P_i^*)\right]_{Pa}, if\ P_i^* \| (\diamondsuit(P_{j1}^*, P_{j2}^*)) \tag{9.5b}$$

$$\begin{aligned}\left[t^2(\| \diamondsuit(P_1^*, P_2^*))\right]_{bi} \cap \left[t^2(\| \diamondsuit(P_{j1}^*, P_{j2}^*))\right] &= \left[t^2(\| \diamondsuit P_{i1}^*, P_{i2}^*)\right]_{Pa}, \\ &if\ (\diamondsuit(P_{i1}^*, P_{i2}^*)) \| (\diamondsuit(P_{j1}^*, P_{j2}^*))\end{aligned} \tag{9.5c}$$

$$\left[t^1(\| P_i^*)\right]_{bi} \cap \left[t^3\right]_{bj} = \left[t^1(\| P_i^*)\right]_{Pa} \tag{9.5d}$$

$$\left[t^2(\| \diamondsuit(P_{i1}^*, P_{i2}^*))\right]_{bi} \cap \left[t^3\right]_{bj} = \left[t^2(\| \diamondsuit(P_{i1}^*, P_{i2}^*))\right]_{Pa} \tag{9.5e}$$

$$\left[t^3\right]_{bi} \cap \left[t^3\right]_{bj} = \left[t^3\right]_{Pa} \tag{9.5f}$$

Obviously, if the geometrical condition on right side of each equation (Eqs. 9.5a–9.5f) is met, no translational element will be eliminated.

It should be noted that if POC set of the sub-PM formed by the first k branches has become the same as the PM's POC set, the (k + 1)th branch can be replaced by a simpler branch. But POC set of this simpler branch should contain at least all elements of the PM's POC set. For example, $SOC\{-S-S-R-\}$ is a branch for this purpose.

Example 9.4 Determine the geometrical conditions for No. 1 branch combination scheme in Table 9.3.

(1) Topological structure of the SOC branch
Topological structure of branches:

$$SOC\{-R_{j1} \parallel R_{j2} \parallel C_{j3}-\}, \mathrm{j} = 1, 2, 3.$$

(2) Selection an arbitrary point o' on the movable platform as the base point.

(3) Determine POC set of the branch
According to Eq. (4.4) in Chap. 4, POC set of the branch is

$$M_{b_j} = \begin{bmatrix} t^3 \\ r^1(\parallel R_{j1}) \end{bmatrix}, \ \mathrm{j} = 1, 2, 3.$$

(4) Establish POC equation for the parallel mechanism
Substitute POC set of the PM (three translations) and POC set of each branch into the POC set for parallel mechanisms (Eq. (5.3) in Chap. 5), there is

$$\begin{bmatrix} t^3 \\ r^0 \end{bmatrix} \Leftarrow \begin{bmatrix} t^3 \\ r^1(\parallel R_{11}) \end{bmatrix} \cap \begin{bmatrix} t^3 \\ r^1(\parallel R_{21}) \end{bmatrix} \cap \begin{bmatrix} t^3 \\ r^1(\parallel R_{31}) \end{bmatrix}$$

where, "$\Leftarrow$" means that POC set on the left side is "intersection" of the POC sets on the right side.

(5) Determine geometrical conditions for assembling of the first two branches.
In order to achieve the PM's POC set, "intersection" of the POC sets of the first two branches should eliminate the rotational element. According to topological structure ($R_{11} \# R_{21}$) and Eq. (9.2a), the above equation can be rewritten as

$$\begin{bmatrix} t^3 \\ r^0 \end{bmatrix} \Leftarrow \begin{bmatrix} t^3 \\ r^0 \end{bmatrix} \cap \begin{bmatrix} t^3 \\ r^1(\parallel R_{31}) \end{bmatrix}$$

where, the first item on right side of the equation is POC set of the sub-PM formed by the first two branches. It is already equal to the PM's POC set.

(6) Determine geometrical conditions for assembling of the third branch.
Since the first two branches have guaranteed three translations and no rotation of the movable platform, the third branch can be assembled arbitrarily.

(7) Geometrical conditions for assembling of this branch combination scheme
The geometrical conditions for branches to be assembled between two platforms are:

(a) Geometrical condition for assembling of the first two branches: $R_{11} \# R_{21}$.

(b) Geometrical condition for assembling of the third branch: can be assembled arbitrarily.

9.5.2 Assembling Schemes of the Branch Combination

When the geometrical conditions for branch combination assembling is determined, there are still several possible assembling schemes. Each possible assembling scheme complies with the determined geometrical conditions.

Example 9.5 For the geometrical conditions for branch assembling determined in Example 9.4, determine different assembling schemes for this branch combination.

According to Example 9.4, geometrical condition for assembling of the first two branches is $R_{11} \# R_{21}$. The third branch can be assembled arbitrarily. There are the following different assembling schemes:

(1) Axes of R_{11}, R_{21} and R_{31} are skew in space, as shown in Fig. 9.5a. Translations of the three C pairs can be used as driving inputs simultaneously (refer to Sect. 6.3.2 in Chap. 6).

(2) Axes of R_{11}, R_{21} and R_{31} are orthogonal in space, as shown in Fig. 9.5b.

(3) Axes of R_{11}, R_{21} and R_{31} form a triangle, as shown in Fig. 9.5c. Translations of the three C pairs cannot be used as driving inputs simultaneously (refer to Sect. 6.3.2 in Chap. 6).

(4) Axes of R_{11}, R_{21} and R_{31} form a letter "Y", as shown in Fig. 9.5d.

(5) Axes of R_{11} and R_{21} are perpendicular and axes of R_{21} and R_{31} are parallel, as shown in Fig. 9.5e. This PM has partial motion decoupling property (refer to the chap. 7 for detail).

(6) Axes of R_{11} and R_{21} are perpendicularly intersected. The third branch is replaced by $SOC\{-S - P - S-\}$, as shown in Fig. 9.5f.

Generally, for the same geometrical assembling condition, one branch combination may have several different assembling schemes. So, the solution space of the PM structure synthesis can be greatly enlarged.

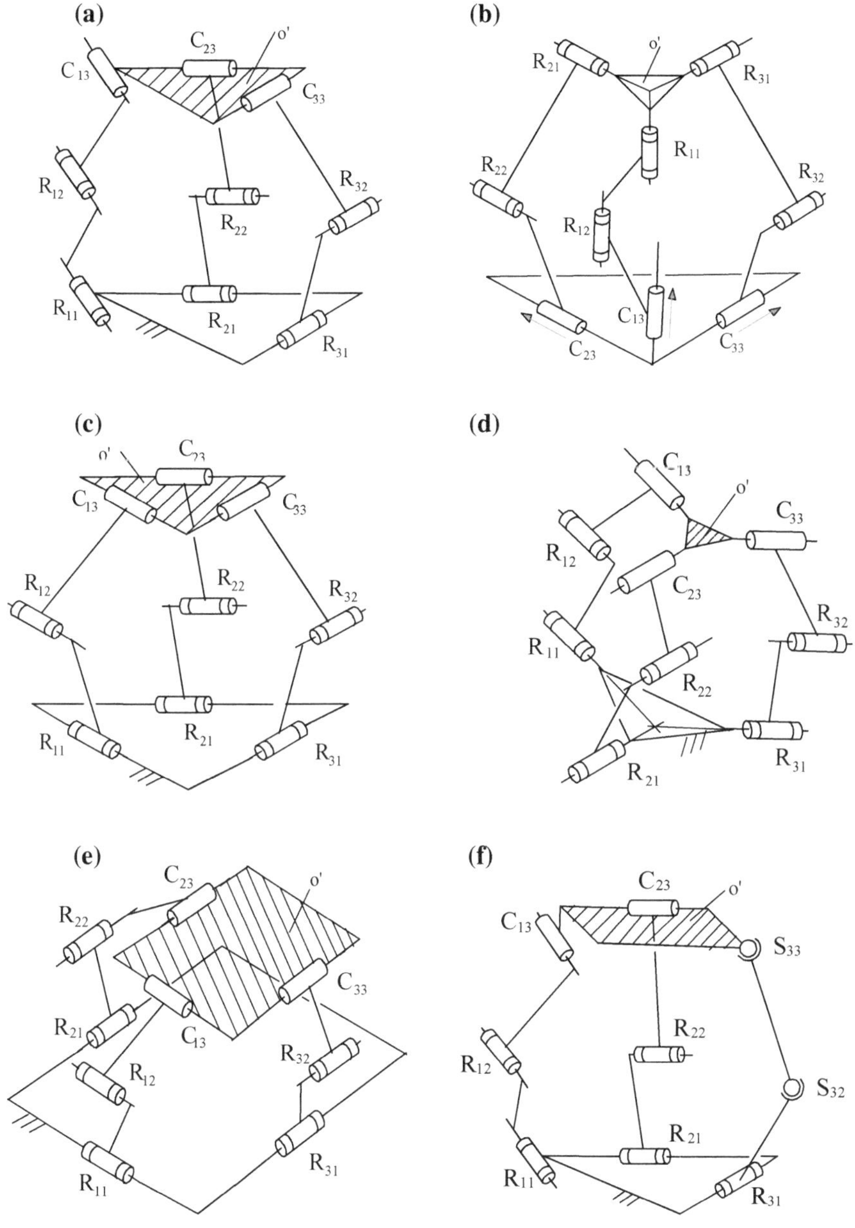

Fig. 9.5 Six different assembling schemes for one branch combination

9.5.3 Check DOF of the PM

For each PM structure type obtained in Example 9.5, its DOF should be checked.

Example 9.6 Check DOF of the mechanism in Fig. 9.5a.

Check DOF of the mechanism in Fig. 9.5a according to DOF calculation method described in Sect. 6.4 in Chap. 6.

(1) According to DOF formula (Eq. 6.8b in Chap. 6), the number of independent displacement equations of the first independent loop is

$$\begin{aligned}\xi_{L_1} &= \dim\{M_{b_1} \cup M_{b_2}\}\\ &= \dim\left\{\begin{bmatrix} t^3 \\ r^1(\| R_{11}) \end{bmatrix} \cup \begin{bmatrix} t^3 \\ r^1(\| R_{21}) \end{bmatrix}\right\} = \dim\left\{\begin{bmatrix} t^3 \\ r^2(\| \diamondsuit(R_{11}, R_{21})) \end{bmatrix}\right\} = 5\end{aligned}$$

(2) According to DOF formula (Eq. 6.8b), the number of independent displacement equations of the second independent loop is

$$\begin{aligned}\xi_{L_2} &= \dim\{(M_{b_1} \cap M_{b_2}) \cup M_{b_3}\}\\ &= \dim\left\{\left(\begin{bmatrix} t^3 \\ r^1(\| R_{11}) \end{bmatrix} \cap \begin{bmatrix} t^3 \\ r^1(\| R_{21}) \end{bmatrix}\right) \cup \begin{bmatrix} t^3 \\ r^1(\| R_{31}) \end{bmatrix}\right\} = \dim\left\{\begin{bmatrix} t^3 \\ r^1(\| R_{31}) \end{bmatrix}\right\} = 4\end{aligned}$$

(3) According to DOF formula (Eq. 6.8a in Chap. 6), DOF of the mechanism in Fig. 9.5a is

$$F = \sum_{i=1}^{m} f_i - \sum_{j=1}^{2} \xi_{L_j} = 12 - (5 + 4) = 3$$

It complies with the design requirement.

Similarly, DOF of the mechanisms in Fig. 9.5b–f can be checked. All these mechanisms comply with the design requirement.

9.5.4 Selection of Driving Pairs

For the obtained PM structure type, check whether location of the driving pairs comply with design requirement.

Example 9.7 Determine whether R_{11}, R_{21} and R_{31} of the mechanism in Fig. 9.5a can be used as driving pairs simultaneously.

(1) According to criteria for driving pairs selection (Sect. 6.3.2 in Chap. 6), select the three R pairs (R_{11}, R_{21} and R_{31}) on the fixed platform as the driving pairs and lock them. A new PM is obtained. Topological structure of this new PM's branch is

$$SOC\{-R_{j2} \parallel C_{j3}-\}, \mathrm{j} = 1, 2, 3.$$

(2) According to POC equation for serial mechanisms (Eq 4.4 in Chap. 4), POC set of the branch is

$$M_{bj} = \begin{bmatrix} t^3 \\ r^1(\parallel R_{j2}) \end{bmatrix}, \mathrm{j} = 1, 2, 3.$$

(3) Determine DOF of the new PM.
According to DOF formula (Eq. 6.8b in Chap. 6), the number of the independent displacement equations of the two independent loops are

$$\begin{aligned}\xi_{L_1} &= \dim\{M_{b_1} \cup M_{b_2}\} \\ &= \dim\{\begin{bmatrix} t^3 \\ r^1(\parallel R_{12}) \end{bmatrix} \cup \begin{bmatrix} t^3 \\ r^1(\parallel R_{22}) \end{bmatrix}\} = \dim\{\begin{bmatrix} t^3 \\ r^2(\parallel \diamondsuit(R_{12}, R_{22})) \end{bmatrix}\} = 5\end{aligned}$$

$$\begin{aligned}\xi_{L_2} &= \dim\{(M_{b_1} \cap M_{b_2}) \cup M_{b_3}\} \\ &= \dim\{(\begin{bmatrix} t^3 \\ r^1(\parallel R_{12}) \end{bmatrix} \cap \begin{bmatrix} t^3 \\ r^1(\parallel R_{22}) \end{bmatrix}) \cup \begin{bmatrix} t^3 \\ r^1(\parallel R_{32}) \end{bmatrix}\} = \dim\{\begin{bmatrix} t^3 \\ r^1(\parallel R_{32}) \end{bmatrix}\} = 4\end{aligned}$$

According to DOF formula (Eq. 6.8a in Chap. 6), DOF of the new PM is

$$F^* = \sum_{i=1}^{m} f_i - \sum_{j=1}^{v} \xi_{L_j} = 9 - (5 + 4) = 0$$

(4) Since DOF of the new PM is 0, all the three R pairs (R_{11}, R_{21} and R_{31}) on the fixed platform can be selected as the driving pairs simultaneously.

Similarly, it can be determined that all the three R pairs (R_{11}, R_{21} and R_{31}) on the fixed platform of the mechanisms in Fig. 9.5b–f can be selected as the driving pairs simultaneously.

9.6 Performance Analysis and Classification of PM Structure Types

9.6.1 Topological Characteristics of Parallel Mechanisms

Based on POC equation for serial mechanisms (Eq. 4.4 in Chap. 4), POC equation for parallel mechanisms (Eq. 5.3 in Chap. 5), DOF formula (Eq. 6.8 in Chap. 6) and AKC coupling degree formula (Eq. 7.9 in Chap. 7), 12 topological characteristics of parallel mechanisms and corresponding calculation formulas (or criteria) are listed in Table 9.4. Among these 12 topological characteristics, the most important are:

Table 9.4 Basic parameters and topological characteristics of parallel mechanisms

Basic parameters		
1	Number of pairs, m	Determined by topological structure
2	Number of links, n	Determined by topological structure
3	Number of branches, n_b	Determined by topological structure
4	Number of SOC branches, $n_{b(SOC)}$	Determined by topological structure
5	Number of HSOC branches, $n_{b(HSOC)}$	Determined by topological structure
6	Number f independent loops, v	$v = m - n + 1$
Topological characteristics		
1	POC set, M_{pa}	Equation (5.3)
2	Dimension of the POC set, $\dim\{M_{pa}\}$	Equation (3.10)
3	DOF, F	Equation (6.8)
4	Number of independent displacement equations, ξ_{Pa}	Equation (6.10)
5	Number of overconstraints, N_{ov}	Equation (6.11)
6	Redundancy, D_{re}	Equation (6.12)
7	Inactive kinematic pair	Criteria (Sect. 6.3.1)
8	Driving pairs	Criteria (Sect. 6.3.2)
9	Type and number of AKC	Criteria (Sect. 7.4)
10	Coupling degree of AKC, κ	Equation (7.9)
11	Type of DOF	Criteria (Sect. 7.5)
12	Motion decoupling property	Criteria (Sect. 7.6)

(1) POC set and DOF.

POC set and DOF are the most important two characteristics. Six other characteristics are derived from them: i.e. dimension of the POC set, number of independent displacement equations, number of overconstraints, redundant DOF, inactive kinematic pair, driving pairs, etc. All these eight topological characteristics have close relations with basic functions and kinematic (dynamic, control) performance of the parallel mechanism.

(2) Type of AKC and coupling degree of AKC.

Type of AKC and coupling degree of AKC are also very important. Two other characteristics are derived from them: i.e. Type of DOF and motion decoupling property. All these four topological characteristics have close relations with kinematic (dynamic, control) performance and complexity of the parallel mechanism.

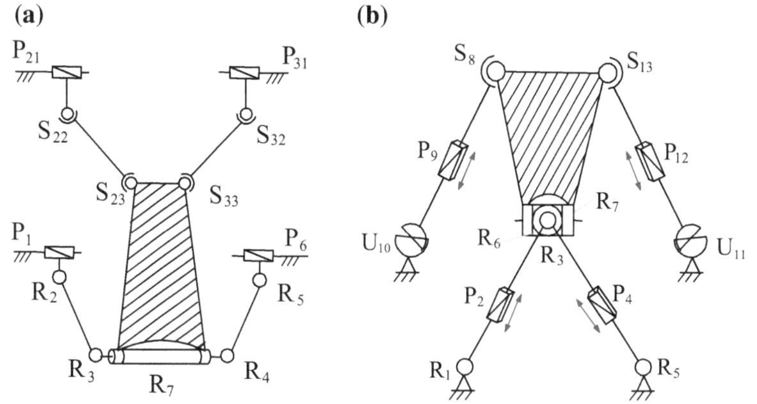

Fig. 9.6 Two (2T-2R) parallel mechanisms

9.6.2 *Performance Analysis of Structure Type*

Based on the topological characteristics of parallel mechanisms in Table 9.4, performance of obtained structure types can be analyzed and compared. The result can be used for classification and optimum selection of PM structure types.

Example 9.8 Determine topological characteristics of the (2T-2R) parallel mechanism in Fig. 9.6a.

(1) Topological structure

- HSOC branch
  The HSOC branch is composed of a planar six-bar linkage (SLC{-P_1-R_2-R_3-R_4-R_5-P_6-}) and a serial connected R_7, as shown in Fig. 9.6a.

- SOC branch

$$SOC\{-P_{i1} - S_{i2} - S_{i3}-\}, (\mathrm{i} = 2, 3).$$

(2) Select an arbitrary point o' on axis of R_7 on the movable platform as the base point.

(3) POC sets of the branches

(a) POC set of the HSOC branch
Substitute POC set of the planar six-bar linkage (No. 7 in Table 9.2) and POC set of R_7 (Eq. 3.2 in Chap. 3) into POC equation for serial mechanisms (Eq. 4.4 in Chap. 4) and there is

$$M_{b_1} = \begin{bmatrix} t^2(\perp R_3) \\ r^1(\parallel R_3) \end{bmatrix} \cup \begin{bmatrix} t^0 \\ r^1(\parallel R_7) \end{bmatrix} = \begin{bmatrix} t^2(\perp R_3) \\ r^2(\parallel \diamondsuit(R_3, R_7)) \end{bmatrix}$$

(b) POC sets of SOC branches
According to POC equation for serial mechanisms, there is

$$M_{b_i} = \begin{bmatrix} t^3 \\ r^3 \end{bmatrix}, \ \mathrm{i} = 2, 3.$$

(4) POC set of the PM
Substitute POC sets of the above three branches into POC equation for parallel mechanisms (Eq. 5.3 in Chap. 5) and there is

$$M_{Pa} = \begin{bmatrix} t^2(\perp R_3) \\ r^2(\| \diamondsuit(R_3, R_7)) \end{bmatrix} \cap \begin{bmatrix} t^3 \\ r^3 \end{bmatrix} \cap \begin{bmatrix} t^3 \\ r^3 \end{bmatrix} = \begin{bmatrix} t^2(\perp R_3) \\ r^2(\| \diamondsuit(R_3, R_7)) \end{bmatrix}$$

(5) DOF of the mechanism

① The first loop is a planar six-bar linkage. The number of independent displacement equations is

$$\xi_{L_1} = 3$$

② According to DOF formula (Eq. 6.8b in Chap. 6), the number of independent displacement equations of the second and the third independent loop is

$$\xi_{L_2} = \xi_{L_3} = 6$$

③ According to DOF formula (Eq. 6.8a), DOF of the mechanism in Fig. 9.6a is

$$F = \sum_{i=1}^{m} f_i - \sum_{j=1}^{2} \xi_{L_j} = 21 - (3 + 6 + 6) = 6$$

It is easy to find that this mechanism has two passive DOFs. The actual DOF of the mechanism is DOF = 4.

According to criteria for inactive pairs determination (Sect. 6.3.1 in Chap. 6), it is easy to determine that this mechanism contains no inactive pairs (omitted here).

(6) Dimension the PM's POC set
Since POC set and DOF of the PM are known, according to Eq. (3.10) in Chap. 3, dimension of the POC set is

$$\dim\{M_{Pa}\} = 4$$

(7) Number of independent displacement equations
According to Eq. (6.10) in Chap. 6, the number of independent displacement equations is

$$\xi_{mech} = \sum_{j=1}^{3} (\dim\{(\bigcap_{i=1}^{j} M_{b_i}) \bigcup M_{b_{(j+1)}}\}) = 3+6+6 = 15$$

(8) Number of overconstraints
According to Eq. (6.11) in Chap. 6, the number of overconstraints is

$$N_{ov.} = 6v - \xi_{mech} = 18 - 15 = 3$$

(9) Redundancy
According to Eq. (6.12) in Chap. 6, redundancy of the mechanism is

$$D_{red.} = F - \dim\{M_{Pa}\} = 4 - 4 = 0$$

(10) Driving pairs
According to criteria for driving pairs selection (Sect. 6.3.2 in Chap. 6), it is easy to determine that P_1, P_6, P_{21} and P_{31} can be used as driving pairs simultaneously (omitted here).

(11) AKC and coupling degree
Determine AKC contained in the mechanism shown in Fig. 9.6a and its coupling degree using the method for topological structure decomposition described in Sect. 7.4 (Chap. 7).

① Determine SOC_1
According to the planar six-bar loop, the first SOC is selected as
$SOC_1\{-P_1 \perp R_2 \parallel R_3 \parallel R_4 \parallel R_5 \perp P_6-\}$, ($P_1$ and P_6 are driving pairs)

② Determine constraint degree of SOC_1
Since $\xi_{L_1} = 3$, according to Eq. (7.6) in Chap. 7, there is

$$\Delta_1 = \sum_{i=1}^{m_1} f_i - I_1 - \xi_{L_1} = 6 - 2 - 3 = +1$$

③ Determine SOC_2
According to the method for topological structure decomposition (Sect. 7.4 in Chap. 7), the second SOC is

$SOC_2\{-R_7 - S_{23} - S_{22} - P_{21}-\}$, ($P_{21}$ is a driving pair)

④ Determine constraint degree of SOC_2
Since $\xi_{L_2} = 6$, according to Eq. (7.6) in Chap. 7, there is

$$\Delta_2 = \sum_{i=1}^{m_2} f_i - I_2 - \xi_{L_2} = 7 - 1 - 6 = 0$$

Where passive DOF is not included.

⑤ Determine SOC_3
According to the method for topological structure decomposition (Sect. 7.4 in Chap. 7), the third SOC is

$$SOC_3\{-S_{33} - S_{32} - P_{31}-\}, \text{ (}P_{31}\text{ is a driving pair)}$$

⑥ Determine constraint degree of SOC_3

Since $\xi_{L_2} = 6$, according to Eq. (7.6) in Chap. 7, there is

$$\Delta_3 = \sum\nolimits_{i=1}^{m_2} f_i - I_2 - \xi_{L_2} = 6 - 1 - 6 = -1$$

Where passive DOF is not included.

⑦ Determine AKC contained in the mechanism and its coupling degree
According to criteria for AKC determination (Sect. 7.4) and Eq. (7.9) in Chap. 7, this PM contains only one $AKC[v = 3, \kappa = 1]$. Coupling degree of this AKC is

$$k = \frac{1}{2}\min\{\sum_{j=1}^{v} |\Delta_j|\} = \frac{1}{2}(|+1|+0+|-1|) = 1$$

Since $\kappa = 1$, forward displacement analysis of this mechanism is relatively easy (refer to Chap. 7).

(12) Type of DOF
Since the PM contains only on AKC, it has full DOF according to method for DOF type determination (Sect. 7.5 in Chap. 7).

(13) Motion decoupling property

Since this PM has full DOF, it has no motion decoupling property according to motion decoupling principle described in Sect. 7.6 in Chap. 7.

For convenience of analysis and comparison, topological characteristics of the PM in Fig. 9.6a are listed in Table 9.5.

Generally, structure synthesis of parallel mechanisms may lead to many structure types. All these obtained structure types can be analyzed following the process of Example 9.8. Topological characteristics of these structure types can be listed in a table (refer to Table 10.4 in Chap. 10).

9.6.3 Classification of Parallel Mechanisms Based on Topological Characteristics

Based on performance analysis, all obtained structure types are compared and classified according to their topological and kinematic performance.

Example 9.9 Determine topological characteristics of the PM in Fig. 9.6b and compared them with those of the PM in Fig. 9.6a.

Topological characteristics of the PM in Fig. 9.6b can be determined following the steps in Example 9.8, as shown in Table 9.5.

As listed in Table 9.5, the two PMs have the following differences:

(1) The PM in Fig. 9.6a contains an HSOC formed by a planar six-bar linkage. The PM in Fig. 9.6b contains an HSOC formed by a planar five-bar linkage.

(2) The PM in Fig. 9.6a contains only one AKC whose coupling degree is $\kappa = 1$. Its forward displacement analysis is relatively simple (refer to Chap. 7). The PM in Fig. 9.6b contains two AKCs (refer to Sect. 7.4 in Chap. 7) whose coupling degree are $\kappa = 1$ and $\kappa = 0$ respectively. Its forward displacement analysis is relatively simple.

(3) The PM in Fig. 9.6a has no motion decoupling. The PM in Fig. 9.6b has partial motion decoupling (refer to Sect. 7.6 in Chap. 7).

Based on the above analysis and comparison, the most suitable structure type can be selected as per design requirement.

Generally, structure synthesis of parallel mechanisms may lead to many structure types. Performance analysis and classification of these structure types can provide useful information for structure assessment and optimization (refer to Sect. 10.2.7 in Chap. 10).

Table 9.5 Topological characteristics of two (2T-2R) PMs

Topological characteristics		The PM in Fig. 9.6a	The PM in Fig. 9.6b
Basic parameters			
1	Number of pairs, m	m = 13	
2	Number of links, n	n = 11	
3	Number of branches, n_b	$n_b = 3$	
4	Number of SOC branches, $n_{b(SOC)}$	$n_{b(SOC)} = 2$	
5	Number of HSOC branches, $n_{b(HSOC)}$	$n_{b(HSOC)} = 1$	
6	Number f independent loops, v	$v = 3$	
Topological characteristics			
1	POC set, M_{pa}	$M_{Pa} = \begin{bmatrix} t^2(\perp R_1) \\ r^2(\parallel \diamond(R_6, R_7)) \end{bmatrix}$	
2	Dimension of the POC set, $\dim\{M_{pa}\}$	$\dim\{M_{pa}\} = 4$	
3	DOF, F	F = 4 (excluding passive DOF)	
4	Number of independent displacement equations, ξ_{Pa}	$\xi_{mech} = 15$	
5	Number of overconstraints, N_{ov}	$N_{ov} = 3$	
6	Redundancy, D_{re}	$D_{re} = \mathbf{0}$	
7	Inactive kinematic pair	No inactive pairs	
8	Driving pairs	Four P pairs	
9	AKC and the coupling degree, κ	$AKC[v = 3, \kappa = 1]$	$AKC_1[v_1 = 1, \kappa_1 = 0]$; $AKC_2[v_2 = 2, \kappa_2 = 1]$
10	Type of DOF	Full DOF	Partial DOF
11	Motion decoupling property	No decoupling	Partial decoupling
12	Structure symmetry	Partial symmetry	

9.7 Summary

(1) A general method for topology design of parallel mechanisms is proposed. The design process includes two stages. The first stage involves structure synthesis in which many different structure types can be obtained (refer to Sect. 10.2.5 in Chap. 10). The second stage includes performance analysis, classification and optimization of the obtained PM structure types (refer to Sects. 10.2.6 and 10.2.7 in Chap. 10).

(2) Structure synthesis method for SOC branches based on POC equation for serial mechanisms (Sect. 9.3.2) and structure synthesis method for HSOC branches based on topological equivalence principle (Sect. 9.3.3) are proposed. Many feasible branch structure types can be obtained.

(3) Basic principle for determining number of branches is proposed. Branch combination schemes are generated based on this principle (Sect. 9.4).

(4) Basic formulas for determining geometrical conditions of assembling branches (Eqs. 9.2a–9.5f) are proposed. For the same geometrical condition, there may be several different assembling schemes (Sect. 9.5.2).

(5) Topological characteristics of PMs are proposed (Table 9.4). These topological characteristics have close relations with the basic function and kinematic (dynamic, control) performance of the mechanism. PM structure types can be analyzed and classified based on these topological characteristics (Sects. 9.6.2 and 9.6.3).

(6) Theoretical bases of this method include four important equations, i.e. POC equation for serial mechanisms (Eq. 4.4 in Chap. 4), POC equation for PMs (Eq. 5.3 in Chap. 5), DOF formula (Eq. 6.8 in Chap. 6) and AKC coupling degree formula (Eq. 7.9 in Chap. 7). Each step in the topology design procedure has explicit formula or criteria. So, this method is easy to understand and easy to use. The complete process of PM topology design will be discussed in Chap. 10.

(7) Since this method is independent of motion position of PM, it can guarantee to obtain the full-cycle DOF PMs. And since it is independent of the fixed coordinate system (i.e. it is not necessary to establish the fixed coordinate system), the geometry conditions of PM existence have generality, such as the PM in Fig. 9.5a, where three axes of R_{11}, R_{21} and R_{31} pairs are skew arbitrarily in space.

(8) Since this method is independent of motion position of PM and the fixed coordinate system, it is called a geometrical method, which is totally different from the other three methods.

References

1. Hunt KH (1978) Kinematic geometry of mechanisms. Clarendon Press, Oxford
2. Frisoli A, Checcacci F et al (2000) Synthesis by screw algebra of translating in-parallel actuated mechanisms. Advance in Robot Kinematics, pp 433–440
3. Huang Z, Li QC (2002) General methodology for type synthesis of symmetrical lower-mobility parallel manipulators and several novel manipulators. Int J Robot Res 21(2): 131–145
4. Huang Z, Li QC (2003) Type synthesis of symmetrical lower-mobility parallel mechanisms using the constraint-synthesis method. Int J Robot Res 22:59–79
5. Kong X, Gosselin CM (2004) Type synthesis of 3T1R 4-DOF parallel manipulators based on screw theory. IEEE Trans Robot Autom 20(2):181–190
6. Kong X, Gosselin CM (2007) Type synthesis of parallel mechanisms. Springer Tracts in Advanced Robotics, vol 33
7. Briot S, Bonev IA (2010) Pantopteron-4: a new 3T1R decoupled parallel manipulator for pick-and-place applications. Mech Mach Theory 45(5):707–721
8. Amine S, Masouleh MT, Caro S, Wenger P, Gosselin C (2012) Singularity conditions of 3T1R parallel manipulators with identical limb structures. ASME J Mech Robot 4(1):011011
9. Huang Z, Li QC, Ding HF (2012) Theory of parallel mechanisms. Springer, Dordrecht
10. Meng X, Gao F, Wu S, Ge J (2014) Type synthesis of parallel robotic mechanisms: framework and brief review. Mech Mach Theory 78:177–186
11. Herve JM (1978) Analyse structurelle des mécanismes par groupe des déplacements. Mech Mach Theory 13:437–450
12. Herve JM (1992) Group mathematics and parallel link mechanisms. In: Proceedings of IMACS/SICE international symposium on robotics, mechatronics, and manufacturing systems, International Association for Mathematics and Computers in Simulation (IMACS), Kobe, pp 459–464
13. Hervé JM (1995) Design of parallel manipulators via the displacement group. In: Proceedings of the 9th world congress on the theory of machines and mechanisms, Milan, Italy, pp 2079–2082
14. Fanghella P, Galletti C (1995) Metric relations and displacement groups in mechanism and robot kinematics. ASME J Mech Des 117:470–478
15. Selig J (1996) Geometrical methods in robotics. Springer, New York
16. Herve JM (1999) The lie group of rigid body displacements, a fundamental tool for mechanism design. Mech Mach Theory 34:719–730
17. Li QC, Huang Z, Herve JM (2004) Type synthesis of 3R2T 5-DOF parallel mechanisms using the lie group of displacements. IEEE Trans Robot Autom 20:173–180
18. Li QC, Herve JM (2009) Parallel mechanisms with bifurcation of schonflies motion. IEEE Trans Robot 25(1):158–164
19. Li QC, Herve JM (2010) 1T2R parallel mechanisms without parasitic motion. IEEE Trans Robot 26:401–410
20. Salgado O, Altuzarra O, Petuya V, Hernandez A (2008) Synthesis and design of a novel 3T1R fully-parallel manipulator. ASME J Mech Des 130:042305-1–042305-8
21. Meng J, Liu GF, Li ZX (2007) A geometric theory for analysis and synthesis of sub-6 DOF parallel manipulators. IEEE Trans Robot 23:625–649
22. Li Z, Lou Y, Zhang Y, Liao B, Li Z (2013) Type synthesis, kinematic analysis, and optimal design of a novel class of schonflies-motion parallel manipulators. IEEE Trans Autom Sci Eng 10(3):674–686
23. Company O, Marquet F, Pierrot F (2003) A new high-speed 4-DOF parallel robot synthesis and modeling issues. IEEE Trans Robot Autom 19(3):411–420
24. Gogu G (2004) Structural synthesis of fully-isotropic translational parallel robots via theory of linear transformations. Eur J Mech A Solids 23:1021–1039
25. Gogu G (2007) Structural synthesis of fully-isotropic parallel robots with schonflies motions via theory of linear transformations and evolutionary morphology. Eur J Mech A Solids 26:242–269
26. Gogu G (2008) Structural synthesis of parallel robots: part 1: methodology. Springer, Dordrecht

27. Gogu G (2009) Structural synthesis of parallel robots: part 2: translational topologies with two and three degrees of freedom. Springer, Dordrecht
28. Gogu G (2010) Structural synthesis of parallel robots: part 3: topologies with planar motion of the moving platform. Springer, Dordrecht
29. Gogu G (2012) Structural synthesis of parallel robots: part 4: other topologies with two and three degrees of freedom. Springer, Dordrecht
30. Yang T-L, Jin Q et al (2001) A general method for type synthesis of rank-deficient parallel robot mechanisms based on SOC unit. Mach Sci Technol 20(3):321–325
31. Jin Q, Yang T-L (2001) Position analysis for a class of novel 3-DOF translational parallel robot mechanisms. In: Proceedings of the ASME 27-th design automation conference, Pittsburgh, ASME International DETC2001/DAC-21151
32. Yang T-L, Jin Q et al (2001) Structural synthesis of 4-DOF (3-translational and 1-rotation) parallel robot mechanisms based on the unites of single-opened-chain. In: The 27th design automation conference, Pittsburgh, ASME International DETC2001/DAC-21152
33. Jin Q, Yang T-L (2001) Structure synthesis of a class five-DOF parallel robot mechanisms based on single-opened-chain units. In: The 27th design automation conference, Pittsburgh, ASME International DETC2001/DAC-21153
34. Yang T-L, Jin Q, Liu A-X, Shen H-P, Luo Y-F (2002) Structural synthesis and classification of the 3-DOF translation parallel robot mechanisms based on the unites of single-open chain. Chin J Mech Eng 38(8):31–36. doi:10.3901/JME.2002.08.031
35. Jin Q, Yang T-L (2002) Structure synthesis and analysis of parallel manipulators with 2-dimension translation and 1-dimension rotation. In: Proceedings of ASME 28th design automation conference, Montreal, DETC 2002/MECH 34307
36. Jin Q, Yang T-L (2002) Synthesis and analysis of a group of 3 DOF (1T-2R) decoupled parallel manipulator. In: Proceedings of ASME 28th design automation conference, Montreal, DETC 2002/MECH 34240
37. Jin Q, Yang T-L (2004) Theory for topology synthesis of parallel manipulators and its application to three-dimension-translation parallel manipulators. ASME J Mech Des 126:625–639
38. Yang T-L (2004) Theory of topological structure for robot mechanisms. China Machine Press, Beijing
39. Jin Q, Yang T-L (2004) Synthesis and analysis of a group of 3-degree-of-freedom partially decoupled parallel manipulators. ASME J Mech Des 126:301–306
40. Shen H-P, Yang T-L et al (2005) Synthesis and analysis of kinematic structures of 6-DOF parallel robotic mechanisms. Mech Mach Theory 40:1164–1180
41. Shen H-P, Yang T-L et al (2005) Structure and displacement analysis of a novel three-translation parallel mechanism. Mech Mach Theory 40:1181–1194
42. Yang T-L, Liu A-X, Luo Y-F et al (2009) Position and orientation characteristic equation for topological design of robot mechanisms. ASME J Mech Des 131:021001-1–021001-17
43. Yang T-L, Liu A-X, Shen H-P et al (2013) On the correctness and strictness of the POC equation for topological structure design of robot mechanisms. ASME J Mech Robot 5(2): 021009-1–021009-18
44. Yang T-L, Sun D-J (2006) General formula of degree of freedom for parallel mechanisms and its application. In: Proceedings of ASME 2006 mechanisms conference, DETC2006-99129
45. Yang T-L, Sun D-J (2008) Rank and mobility of single loop kinematic chains. In: Proceedings of the ASME 32-th mechanisms and robots conference, DETC2008-49076
46. Yang T-L, Sun D-J (2012) A general DOF formula for parallel mechanisms and multi-loop spatial mechanisms. ASME J Mech Robot 4(1):011001-1–011001-17
47. Yang T-L, Liu A-X, Shen H-P et al (2015) Composition principle based on SOC unit and coupling degree of BKC for general spatial mechanisms. In: The 14th IFToMM world congress, Taipei, Taiwan, 25–30 Oct 2015. doi:10.6567/IFToMM.14TH.WC.OS13.135
48. Yang T-L, Shen H-P, Liu A-X, Dai JS (2015) Review of the formulas for degrees of freedom in the past ten years. J Mech Eng 51(3):69–80
49. Yang T-L, Liu A-X, Shen H-P et al (2012) Theory and application of robot mechanism topology. China Science press, Beijing
50. Yang T-L, Liu A-X, Luo Y-F, Shen H-P et al (2010) Basic principles, main characteristics and development tendency of methods for robot mechanism structure synthesis. J Mech Eng 46(9):1–11

Chapter 10
Topology Design of (3T-1R) Parallel Mechanisms

10.1 Introduction

The SCARA parallel robot can achieve 1 rotation and 3 translations (also called Schoenflies motion [1]) and has been used widely in industries such as sorting, packaging and assembling.

The SCARA serial robot first developed by Hiroshi has 4 DOFs [2]. It has a cantilever design. So it features low rigidity, low load capacity and high arm inertia. It is hard for its end effector to move at high speed.

The SCARA parallel robot FlexPicker developed by ABB Corporation has one output platform and one sub-platform. It is obtained by adding a RUPU branch, which may bring in one rotation, to the original Delta parallel robot [3–5].

The SCARA parallel robot Par4 developed by Pierrot has one output platform and 2 sub-platforms [6–16]. Relative motion between the two sub-platforms is converted to rotation of the output platform. But a rotation mechanism is added to expand the manipulator's rotation range [11–20]. This robot features high flexibility, high speed and high rigidity.

In order to obtain a larger rotation range without adding a rotation mechanism, several SCARA parallel robots with only one moving platform are proposed [17–21]. For example, the SCARA parallel robot X4 has one output platform and 4 identical parallel-connected branches. Its movable platform has a rotation within the range of ±90° [21].

In the last decade, many researchers discussed the structure synthesis of (3T-1R) parallel mechanisms [2–44]. The methods used can be divided into the following 4 categories: displacement sub-group/sub-manifold based method [22–25, 37], screw theory based method [26–29, 37], the method based on theory of linear transformations and evolutionary morphology [30, 31, 37] and the POC equation based method presented by authors [32–37, 43–45].

T.-L. Yang et al., *Topology Design of Robot Mechanisms*,
https://doi.org/10.1007/978-981-10-5532-4_10

Topology design of (3T-1R) parallel mechanisms is systematically studied in this chapter. The design process includes the following steps:

(1) Basic requirements for topology design.
(2) Structure synthesis of branches.
(3) Determine branch combination schemes.
(4) Determine branch assembling schemes.
(5) List all obtained (3T-1R) PM structure types.
(6) Performance analysis of (3T-1R) PM structure types.
(7) Classification of (3T-1R) PM structure types.

18 structure types of (3T-1R) PM containing no P pair are obtained, 2 of which have already been used in existing SCARA parallel robots. Performance analysis and classification are conducted for these structure types.

10.2 Topology Design of (3T-1R) Parallel Mechanisms

10.2.1 Basic Requirements for Topology Design

(1) POC set of the movable platform is

$$M_{Pa} = \begin{bmatrix} t^3 \\ r^1 \end{bmatrix} \tag{10.1}$$

The movable platform has three translations and one rotation around normal line of the fixed platform.

(2) DOF of the mechanism is DOF = 4. The DOF is equal to dimension of the POC set.

(3) Other requirements: The parallel mechanisms should contain no P pairs and no passive DOF. All driving pairs can be selected on the same platform.

10.2.2 Structure Synthesis of Branches

Step 1 **Basic requirements**

(1) POC set of branches
According to Eqs. (10.1) and (5.4) in Chap. 5 (i.e., $M_{b_j} \supseteq M_{Pa}$), POC set of candidate branch may be

$$M_{b_j} = \begin{bmatrix} t^3 \\ r^1 \end{bmatrix}, \begin{bmatrix} t^3 \\ r^2 \end{bmatrix}, \begin{bmatrix} t^3 \\ r^3 \end{bmatrix}. \tag{10.2}$$

These candidate branches are written as (3T-1R), (3T-2R) or (3T-3R) branches.

(2) DOF of the branch is equal to dimension of its POC set, i.e. $DOF = dim\{M_{b_j}\}$.

Step 2 **Structure syntheses of SOC branches**

(1) Structure synthesis of the SOC branch

Since POC set and DOF of the SOC branch are known, structure types of the SOC branch containing only R and P pairs can be obtained according to the structure synthesis method based on POC equation for serial mechanisms in Chap. 8.

Structure types for SOC branches containing only R and P pairs have been listed in Table 8.1 in Chap. 8. The SOC branch corresponding to the prescribed POC set can be selected directly from Table 8.1. For the three types of POC sets in Eq. (10.2), eight structure types of SOC branch with DOF = dim{M_S} can be selected, as shown in Table 10.1.

Table 10.1 SOC branches and HSOC branches

M_{b_j}	SOC branches	HSOC branches
(3T-1R)	(1) $SOC\{-R\|R\|R-P-\}$	(1) $HSOC\{-R\|R\|(P^{(4R)})\|R-\}$ (Fig. 10.1a) (2) $HSOC\{-R\|R\|(R^{(2S2R)}\|P^{(2S2R)})-\}$ (Fig. 10.1b)
	(2) $SOC\{-R\|R-P-P-\}$	(3) $HSOC\{-R\|(R^{(4S)}\|P^{(4S)}\perp P^{(4S)})-\}$ (Fig. 10.1c) (4) $HSOC\{-R\|R-(\Diamond(P^{(2R1C-3R)},P^{(2R1C-3R)})-\}$ (Fig. 10.1d) (5) $HSOC\{-R(\|\Diamond(P^{(4R)},P^{(4R)})\|R-\}$ (Fig. 10.1e)
	(3) $SOC\{-R-P-P-P-\}$	(6) $HSOC\{-R\perp 2\{R\|(R^{(4S)}\|P^{(4S)}\perp P^{(4S)})-\}-\}$ (Fig. 10.1f) (7) $HSOC\{-R\perp Delta\,PM-\}$ (Fig. 10.1g)
(3T-2R)	(4) $SOC\{-R\|R\|R-R\|R-\}$	No corresponding HSOC branch
	(5) $SOC\{-R\|R\|R-R-P-\}$	(8) $HSOC\{-R\|R\|(P^{(4R)})\|R\perp R-\}$ (Fig. 10.2a) (9) $HSOC\{-R\|R\|(R^{(2S2R)}\|P^{(2S2R)})\perp R-\}$ (Fig. 10.2b)
	(6) $SOC\{-R\|R-R\|R-P-\}$	(10) $HSOC\{-R\|R(-P^{(4R)})\perp R\|R-\}$ (11) $HSOC\{-R\|(R^{(2S2R)}\|P^{(2S2R)})\perp R\|R-\}$
	(7) $SOC\{-R\|R-R-P-P-\}$	(12) $HSOC\{-R(\|R^{(4S)}\|P^{(4S)}\perp P^{(4S)})\perp R-\}$ (Fig. 10.2c) (13) $HSOC\{-\Diamond(P^{(2R1C-3R)},P^{(2R1C-3R)})\|R\|R\perp R-\}$ (Fig. 10.2d)
(3T-3R)	(8) $SOC\{-S-S-R-\}$	No corresponding HSOC branch

(2) Expansion of structure types of the SOC branch

For SOC branches listed in Table 10.1, new SOC branches can be generated using the following methods:

(a) Change connection order of kinematic pairs.
For example, for No. 2 SOC branch in Table 10.1, the following four SOC branches can be obtained by changing connection order of kinematic pairs.

① $SOC\{-R \parallel R - P - P-\}$;

② $SOC\{-R(-P) \parallel R - P-\}$, axes of two R pairs are kept parallel.

③ $SOC\{-R(-P - P) \parallel R-\}$, axes of two R pairs are kept parallel.

④ $SOC\{-P - R \parallel R - P-\}$.

(b) Merge several R pairs and P pairs into a C pair or U pair.
For example, for No. 1 SOC branch in Table 10.1, new SOC branch $SOC\{-R \parallel R \parallel C-\}$ can be obtained by merging the R pair and the P pair into a C pair. For No. 4 SOC branch in Table 10.1, new SOC branch $SOC\{-R \parallel R \parallel \overbrace{R \perp R} \parallel R-\}$ can be obtained by merging the third and the fourth R pair into a U pair ($\overbrace{R \perp R}$ is the symbolic representation of U pair, refer to Chap. 2).

(c) Replace several R pairs with an S pair.
For example, three R pairs whose axes intersect at one common point $\left(SOC\{-\overbrace{RRR}-\}\right)$ can be replaced by an S pair.

Although there are only a finite number of SOC branches containing only R and P pairs (Table 10.1). Much more SOC branches containing multi-DOF kinematic pairs (C, U, S, etc.) can be obtained through expansion.

Step 3 **Structure syntheses of HSOC branches**

According the method for structure synthesis of HSOC branches described in Sect. 9.3.3 in Chap. 9, select an SOC branch with the same POC set and the same DOF as the HSOC to be synthesized and replace a sub-SOC of the SOC branch with a topologically equivalent sub-PM. Then an HSOC branch can be obtained.

A. **Structure synthesis of (3T-1R) HSOC branch**

(1) For $SOC\{-R\|R\|R - P-\}$ (No. 1 SOC branch in Table 10.1), several HSOC branches can be obtained. The following are two such examples:

(a) Replace sub-$SOC\{-P-\}$ of $SOC\{-R\|R\|R - P-\}$ with topologically equivalent sub-$PM\{-P^{(4P)}-\}$ (No. 1 in Table 9.2 in Chap. 9) and $HSOC\{-R\|R\|(P^{(4R)}) \parallel R-\}$ can be obtained as shown in Fig. 10.1a (No. 1 HSOC in Table 10.1).

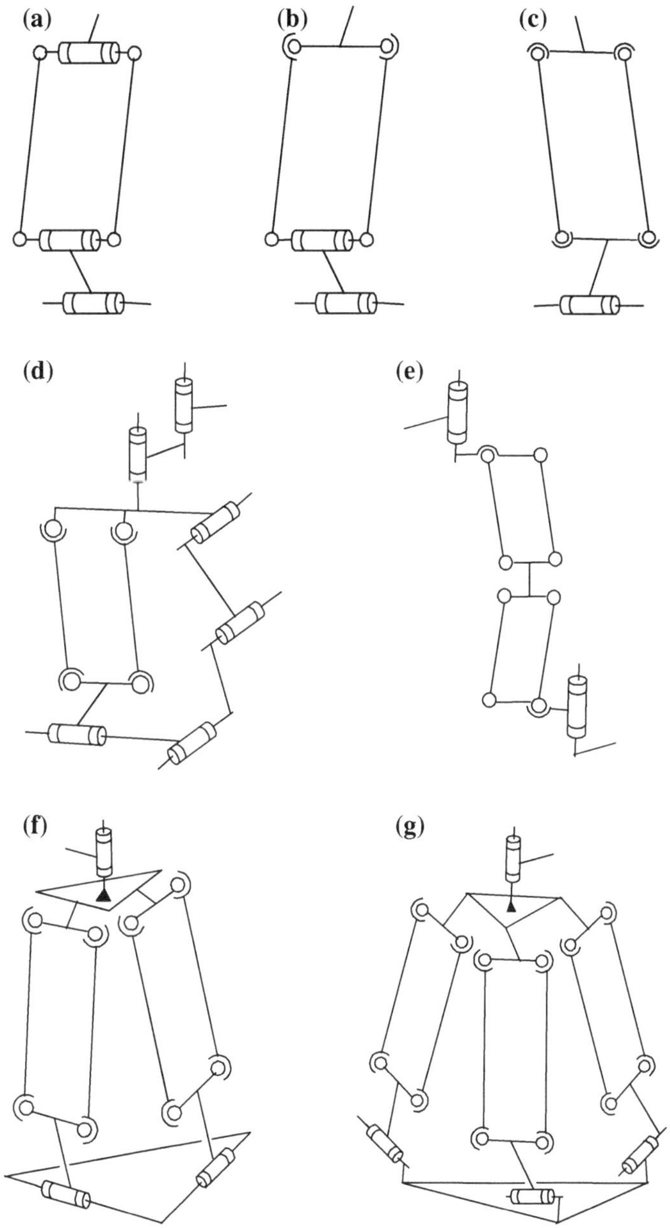

Fig. 10.1 (3T-1R) HSOC branches

(b) Replace sub-$SOC\{-R-P-\}$ of $SOC\{-R\|R\|R-P-\}$ with topologically equivalent sub-$PM\{-R^{(2S2R)}\|P^{(2S2R)}-\}$ (No. 2 in Table 9.2) and $HSOC\{-R\|R\|(R^{(2S2R)}\|P^{(2S2R)})-\}$ can be obtained as shown in Fig. 10.1b (No. 2 HSOC in Table 10.1).

(2) For $SOC\{-R\|R-P-P-\}$ (No. 2 SOC branch in Table 10.1), several HSOC branches can be obtained. The following are three such examples:

(a) Replace sub-$SOC\{-R-P-P-\}$ of $SOC\{-R\|R-P-P-\}$ with topologically equivalent sub-$PM\{-R^{(4S)}\|P^{(4S)}\perp P^{(4S)}-\}$ (No. 3 in Table 9.2) and $HSOC\{-R\|(R^{(4S)}\|P^{(4S)}\perp P^{(4S)})-\}$ can be obtained as shown in Fig. 10.1c (No. 3 HSOC in Table 10.1).

(b) Replace sub-$SOC\{-P-P-\}$ of $SOC\{-R\|R-P-P-\}$ with topologically equivalent sub-$PM\{-\Diamond(P^{(2R1C-3R)},P^{(2R1C-3R)})-\}$ (No. 4 in Table 9.2) and $HSOC\{-R\|(R^{(4S)}\|P^{(4S)}\perp P^{(4S)})-\}$ can be obtained as shown in Fig. 10.1d (No. 4 HSOC in Table 10.1).

(c) Replace two sub-$SOC\{-P-\}$ of $SOC\{-R\|R-P-P-\}$ with topologically equivalent sub-sub-$PM\{-P^{(4P)}-\}$ (No. 1 in Table 9.2) and change connection order of kinematic pairs. $HSOC\{-R(\|\Diamond(P^{(4R)},P^{(4R)})\|R-\}$ can be obtained. $\Diamond(P^{(4R)},P^{(4R)})$ means that the two parallelograms are coplanar and axes of two R pairs are kept parallel, as shown in Fig. 10.1e (No. 5 HSOC in Table 10.1).

(3) For $SOC\{-R-P-P-P-\}$ (No. 3 SOC branch in Table 10.1), several HSOC branches can be obtained. The following are two such examples:

(a) Replace sub-$SOC\{-P-P-P-\}$ of $SOC\{-R-P-P-P-\}$ with topologically equivalent sub- $PM\{-P^{(2-RRC)}-P^{(2-RRC)}-P^{(2-RRC)}-\}$ (No. 5 in Table 9.2) and further replace sub-$SOC\{-R\|R\|C-\}$ (Fig. 9.3a in Chap. 9) of the sub-PM with topologically equivalent $HSOC\{-R\|(R^{(4S)}\|P^{(4S)}\perp P^{(4S)})-\}$ (Fig. 9.3b). Then $HSOC\{-R-2\{R\|(R^{(4S)}\|P^{(4S)}\perp P^{(4S)})-\}-\}$ can be obtained, as shown in Fig. 10.1f (No. 6 HSOC in Table 10.1).

(b) Replace sub-$SOC\{-P-P-P-\}$ of $SOC\{-R-P-P-P-\}$ with topologically equivalent Delta PM and $HSOC\{-R\perp Delta\,PM-\}$ can be obtained as shown in Fig. 10.1g (No. 7 HSOC in Table 10.1).

B. **Structure synthesis of (3T-2R) HSOC branch**

(1) For $SOC\{-R\|R\|R-R-P-\}$ (No. 5 SOC branch in Table 10.1), several HSOC branches can be obtained. The following are two such examples:

(a) Replace sub-$SOC\{-P-\}$ of $SOC\{-R\parallel R\parallel R-R-P-\}$ with topologically equivalent sub-$PM\{-P^{(4P)}-\}$ (No. 1 in Table 9.2) and $HSOC\{-R\parallel R\parallel(P^{(4R)})\parallel R\perp R-\}$ can be obtained as shown in Fig. 10.2b (No. 8 HSOC in Table 10.1).

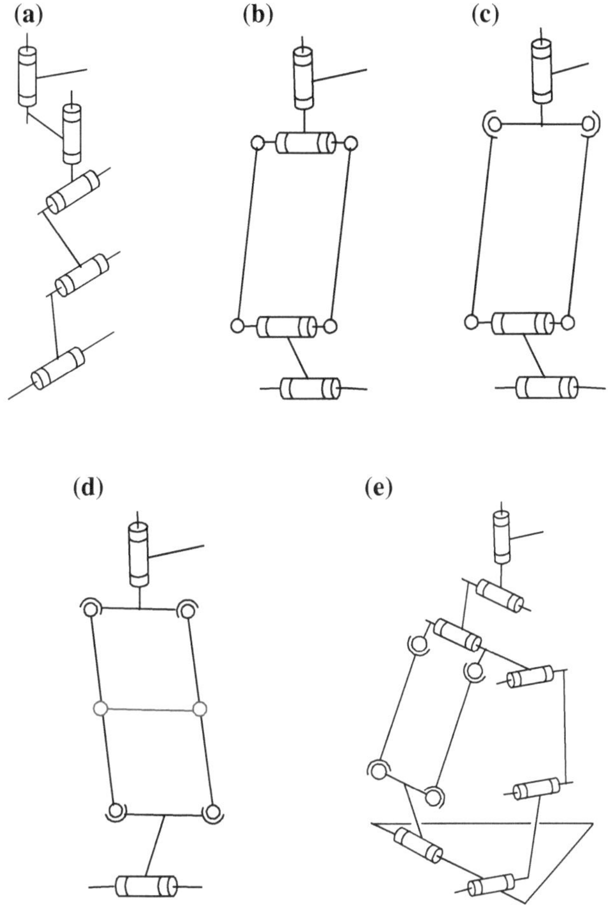

Fig. 10.2 (3T-2R) SOC branch and topologically equivalent HSOC branches

(b) Replace sub-$SOC\{-R-P-\}$ of $SOC\{-R\|R\|R-R-P-\}$ with topologically equivalent sub-$PM\{-R^{(2S2R)}\|P^{(2S2R)}-\}$ (No. 2 in Table 9.2) and $HSOC\{-R\parallel R\parallel (R^{(2S2R)}\parallel P^{(2S2R)})\perp R-\}$ can be obtained as shown in Fig. 10.2c (No. 9 HSOC in Table 10.1).

(2) For $SOC\{-R\|R-R\|R-P-\}$ (No. 6 SOC branch in Table 10.1), several HSOC branches can be obtained. The following are two such examples:

(a) Replace sub-$SOC\{-P-\}$ of $SOC\{-R\parallel R-R\parallel R-P-\}$ with topologically equivalent sub-$PM\{-P^{(4P)}-\}$ (No. 1 in Table 9.2) and $HSOC\{-R\parallel R(-P^{(4R)})\perp R\parallel R-\}$ can be obtained (No. 10 HSOC in Table 10.1.)

(b) Replace sub-$SOC\{-R-P-\}$ of $SOC\{-R\parallel R-R\parallel R-P-\}$ with topologically equivalent sub-$PM\{-R^{(2S2R)}\|P^{(2S2R)}-\}$ (No. 2 in Table 9.2) and $HSOC\{-R\parallel (R^{(2S2R)}\parallel P^{(2S2R)})\perp R\parallel R-\}$ can be obtained (No. 11 HSOC in Table 10.1).

(3) For $SOC\{-R\|R-R-P-P-\}$ (No. 7 SOC branch in Table 10.1), several HSOC branches can be obtained. The following are two such examples:

(a) Replace sub-$SOC\{-R-P-P-\}$ of $SOC\{-R\|R-R-P-P-\}$ with topologically equivalent sub-$PM\{-R^{(4S)}\|P^{(4S)}\perp P^{(4S)}-\}$ (No. 3 in Table 9.2) and $HSOC\{-R\|(R^{(4S)}\|P^{(4S)}\perp P^{(4S)})\perp R-\}$ can be obtained, as shown in Fig. 10.2d (No. 12 HSOC in Table 10.1). In order to avoid folding of the 4S parallelogram along its diagonal, a link with two R pairs is attached.

(b) Replace sub-$SOC\{-P-P-\}$ of $SOC\{-R-R\|R-P-P-\}$ with topologically equivalent sub-$PM\{-\Diamond(P^{(2R1C-3R)},P^{(2R1C-3R)})-\}$ (No. 4 in Table 9.2) and further replace sub-$SOC\{-R\|R\|C-\}$ (Fig. 9.3a) of the sub-PM with topologically equivalent $HSOC\{-R\|(R^{(4S)}\|P^{(4S)}\perp P^{(4S)})-\}$ (Fig. 9.3b). Then $HSOC\{-\Diamond(P^{(1R4S-3R)},P^{(1R4S-3R)})\|R\|R\perp R-\}$ can be obtained, as shown in Fig. 10.2e (No. 13 HSOC in Table 10.1).

Similar to the above synthesis process, HSOC branches containing one driving P pair can also be synthesized.

Although there are only a finite number of SOC branches (Table 10.1), many HSOC branches can be generated by using some sub-PMs in Table 9.2 in Chap. 9. New topologically equivalent HSOC branches lead to new PM structure types.

10.2.3 Branch Combination Schemes

Many structure types of SOC branch and HSOC branch can be obtained during structure synthesis of branches. In order for the PM to achieve the prescribed POC set, the number of branches and branch combination schemes should be determined.

Step 1 **Number of branches**

As discussed in Sect. 9.4 in Chap. 9, the following factors should be considered when determining the number of branches:

(1) According to POC equation for parallel mechanisms (Eq. 5.3 in Chap. 5) and minimum number of branches (Eq. 9.1 in Chap. 9), there are:

(a) For (3T-1R) branches in Table 10.1, at least one branch should be contained. Considering symmetry of the PM, there should be at least two such branches.

(b) For (3T-2R) branches in Table 10.1, there should be at least two such branches.

(2) All driving pairs should be located on the same platform.

(a) If each branch can have only one driving pair, the number of branches should be equal to DOF of the mechanism (DOF = 4).

(b) If HSOC branches are contained, the total number of frame-connected kinematic pairs should be at least equal to DOF of the mechanism (DOF = 4).

(3) The number of branches is equal to the prescribed number of kinematic pairs on the movable platform. For example, the movable platform may sometimes contain only two kinematic pairs [16].

(4) Redundant branch is not considered.

Obviously, the most important factors are structure type of the branch and all driving pair being on the same platform.

Step 2 **Branch combination schemes**

Based on the two SOC branches (SOC-(4) and SOC-(8) in Table 10.1) and 9 HSOC branches (HSOC-(1), (4)-(9), (12), (13) in Table 10.1), different branch combination schemes which can generate the desired POC set and DOF of the PM can be obtained. During branch combination, we shall ensure that the branch shall have simple structure and certain symmetry. Each branch shall have only one driving pair and all driving pairs shall be allocated on the same platform. Table 10.2 shows some of these branch combination schemes.

10.2.4 Branch Assembling Schemes

Step 1 **Geometrical condition for branch assembling**

When branch combination scheme is known, the geometrical condition for assembling branch between two platforms can be determined according to the basic equations (Eqs. 9.2a–9.5f in Chap. 9).

Example 10.1 For the No. 1 branch combination scheme in Table 10.2, determine the corresponding geometrical condition for assembling branches between two platforms.

(1) Topological structure of the four branches:

$$SOC\{-R_{j1}\|R_{j2}\|\widehat{R_{j3}\perp R_{j4}}\|R_{j5}-\}, \mathrm{j}=1,2,3,4.$$

(2) Select an arbitrary point on movable platform as the base point o′ (origin of the moving coordinate system).

Table 10.2 Branch combination schemes of (3T-1R) parallel mechanism

Branch types	Branch combination schemes	
Table 10.1-SOC-(4)	1	$4-SOC\{-R\Vert R\Vert R-R\Vert R-\}$
	2	$2-SOC\{-R\Vert R\Vert R-R\Vert R-\}\oplus 2-SOC\{-S-S-R-\}$
Table 10.1-HSOC-(1)	3	$4-HSOC\{-R\Vert R\Vert(P^{(4R)})\Vert R-\}$
	4	$4-HSOC\{-R\Vert R\Vert(P^{(4R)})\Vert R-\}\oplus 2-SOC\{-S-S-R-\}$
Table 10.1-HSOC-(4)	5	$2-HSOC\{-R(\Vert\Diamond(P^{(4R)},P^{(4R)})\Vert R-\}\oplus 2-SOC\{-S-S-R-\}$
Table 10.1-HSOC-(5)	6	$1-HSOC\{-R\Vert R-(\Diamond(P^{(1R4S-3R)},P^{(1R4S-3R)})-\}\oplus 2-SOC\{-S-S-R-\}$
Table 10.1-HSOC-(6)	7	$2-HSOC\{-R\perp 2\{R\Vert(R^{(4S)}\Vert P^{(4S)}\perp P^{(4S)})-\}-\}$
	8	$1-HSOC\{-R\perp 2\{R\Vert(R^{(4S)}\Vert P^{(4S)}\perp P^{(4S)})-\}-\}\oplus 2-SOC\{-S-S-R-\}$
Table 10.1-HSOC-(7)	9	$1-HSOC\{-R-Delta\,PM-\}\oplus 1-SOC\{-S-S-R-\}$
Table 10.1-HSOC-(8)	10	$4-HSOC\{-R\Vert R\Vert(P^{(4R)})\Vert R\perp R-\}$
	11	$2-HSOC\{-R\Vert R\Vert(P^{(4R)})\Vert R\perp R-\}\oplus 2-SOC\{-S-S-R-\}$
Table 10.1-HSOC-(9)	12	$4-HSOC\{-R\Vert R\Vert(R^{(2S2R)}\Vert P^{(2S2R)})\perp R-\}$
	13	$2-HSOC\{-R\Vert R\Vert(R^{(2S2R)}\Vert P^{(2S2R)})\perp R-\}\oplus 2-SOC\{-S-S-R-\}$
Table 10.1-HSOC-(12)	14	$4-HSOC\{-R(\Vert R^{(4S)}\Vert P^{(4S)}\perp P^{(4S)})\perp R-\}$
	15	$2-HSOC\{-R\Vert(R^{(4S)}\Vert P^{(4S)}\perp P^{(4S)})\perp R-\}\oplus 2-SOC\{-S-S-R-\}$
Table 10.1-HSOC-(13)	16	$2-HSOC\{-\Diamond(P^{(1R4S-3R)},P^{(1R4S-3R)})\Vert R\Vert R\perp R-\}$

(3) Determine POC set of the SOC branch
According to Eq. (4.4) in Chap. 4 or Table 10.1, POC sets of these four SOC branches are

$$M_{bj} = \begin{bmatrix} t^3 \\ r^2(\| \diamondsuit(R_{j3}, R_{j4}) \end{bmatrix}, \mathrm{j} = 1, 2, 3, 4.$$

(4) Establish POC equation for this parallel mechanism
Substitute the desired POC set of the PM and POC set of each branch into Eq. (5.3) in Chap. 5, then there is

$$\begin{bmatrix} t^3 \\ r^1 \end{bmatrix} \Leftarrow \begin{bmatrix} t^3 \\ r^2(\| \diamondsuit(R_{13}, R_{14}) \end{bmatrix} \cap \begin{bmatrix} t^3 \\ r^2(\| \diamondsuit(R_{23}, R_{24}) \end{bmatrix} \cap \begin{bmatrix} t^3 \\ r^2(\| \diamondsuit(R_{33}, R_{34}) \end{bmatrix} \cap \begin{bmatrix} t^3 \\ r^2(\| \diamondsuit(R_{43}, R_{44}) \end{bmatrix}$$

where, "$\Leftarrow$" means the POC set on the left side of the equation is to be obtained by intersection operation of all the POC sets on the right side of the equation.

(5) Determine geometrical conditions for branch assembling

(a) Determine geometrical conditions for assembling the first two branches
In order for the moving platform to obtain the desired POC set, intersection of POC sets of the first two branches must eliminate one rotational element. According to Eq. (9.2c) in Chap. 9, when $\diamondsuit(R_{13}, R_{14})$ is not parallel to $\diamondsuit(R_{23}, R_{24})$, these two planes have an intersection line. The moving platform has a rotation around this intersection line.

(b) Determine geometrical conditions for assembling the other two branches
Since POC set of the sub-PM formed by the first two branches and the two platforms is already identical to the desired POC set of the PM, assembly of the other two branches shall not change POC set of the PM. So, these two branches can be assembled arbitrarily.

(6) Discussion
Since POC set of the sub-PM formed by the first two branches and the two platforms is already identical to the desired POC set of the PM, the other two branches can be replaced by two simpler SOC branches, such as $SOC\{-S-S-R-\}$ branch (POC set of the simpler SOC branch must contain POC set of the PM) (refer to branch combination scheme 2 in Table 10.2).

Step 2 **Branch assembling schemes**

For the geometrical condition for branch assembling determined in above step 1, there may be several different branch assembling schemes.

Example 10.2 For the No. 1 branch combination scheme in Table 10.2, determine the branch assembling schemes.

According to Example 10.1, the geometrical condition for assembling the branches between two platforms is: $\diamondsuit(R_{13}, R_{14})$ is not parallel to $\diamondsuit(R_{23}, R_{24})$ for the first two branches and the other two branches can be assembled between the two platforms arbitrarily.

Corresponding to the above geometrical condition, there may be several different assembling schemes, such as:

(1) Scheme 1

(a) Four branches: $SOC\{-R_{j1}\|R_{j2}\|\overbrace{R_{j3}\perp R_{j4}}\|R_{j5}-\}, \text{j} = 1,2,3,4.$

(b) Assembling scheme: $R_{15}\|R_{25}\|R_{35}\|R_{45}, R_{31}\|R_{11}\perp R_{21}\|R_{41}$, as shown in Fig. 10.3a.

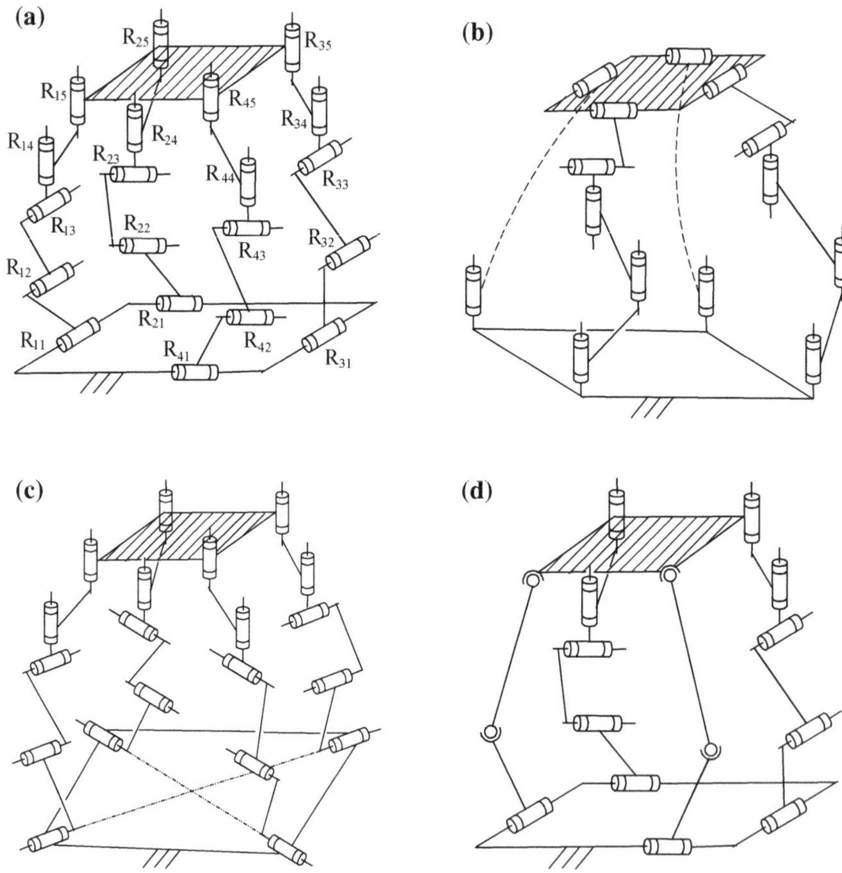

Fig. 10.3 (3T-1R) PMs corresponding to No. 1 and No. 2 branch assembling schemes

(2) Scheme 2

(a) Four branches: $SOC\{-R_{j1}\|R_{j2}\|\overbrace{R_{j3}\bot R_{j4}}\|R_{j5}-\}, \mathrm{j}=1,2,3,4.$

(b) Assembling scheme: $R_{11}\|R_{21}\|R_{31}\|R_{41}, R_{35}\|R_{15}\bot R_{25}\|R_{45}$, as shown in Fig. 10.3b.

(3) Scheme 3

(a) Four branches: $SOC\{-R_{j1}\|R_{j2}\|\overbrace{R_{j3}\bot R_{j4}}\|R_{j5}-\}, \mathrm{j}=1,2,3,4.$

(b) Assembling scheme: $R_{15}\|R_{25}\|R_{35}\|R_{45}, R_{31}|R_{11}\bot R_{21}|R_{41}$, as shown in Fig. 10.3c.

(4) Scheme 4

The first two branches can be assembled in the same way as scheme 1. Then, POC set of the sub-PM formed by the first two branches and the two platforms is already identical to the desired POC set of the PM, the other two branches can be replaced by two $SOC\{-S-S-R-\}$ branches, as shown in Fig. 10.3d.

Obviously, there may be many other assembling schemes and the above four assembling schemes are only several examples.

Step 3 **Check DOF of the mechanism**

For the obtained (3T-1R) parallel mechanism, check whether its DOF meets the design requirement.

Example 10.3 Check whether DOF of the parallel mechanism in Fig. 10.3a meets the design requirement.

(1) Check POC set of the branch
According to Example 10.1, POC set of each branch is

$$M_{bj}=\begin{bmatrix} t^3 \\ r^2(\| \diamondsuit(R_{j3},R_{j4}) \end{bmatrix}, \quad \mathrm{j}=1,2,3,4.$$

(2) Determine the number of independent displacement equations of each independent loop.

According to the DOF formula (Eq. (6.8b) in Chap. 6) and the branch assembling scheme, the number of independent displacement equations of each independent loop is:

$$\xi_{L_1} = \dim\{M_{b_1} \cup M_{b_2}\}$$
$$= \dim\left\{\begin{bmatrix} t^3 \\ r^2(\| \diamond (R_{13}, R_{14})) \end{bmatrix} \bigcup \begin{bmatrix} t^3 \\ r^2(\| \diamond (R_{23}, R_{24})) \end{bmatrix}\right\} = 6$$

$$\xi_{L_2} = \dim\left\{\left(\bigcap_{j=1}^{2} M_{b_j}\right) \bigcup M_{b_3}\right\}$$
$$= \dim\left\{\bigcap_{j=1}^{2}\begin{bmatrix} t^3 \\ r^2(\| \diamond (R_{j3}, R_{j4})) \end{bmatrix} \bigcup \begin{bmatrix} t^3 \\ r^2(\| \diamond (R_{33}, R_{34})) \end{bmatrix}\right\} = 5$$

$$\xi_{L_3} = \dim\left\{\left(\bigcap_{j=1}^{3} M_{b_j}\right) \bigcup M_{b_4}\right\}$$
$$= \dim\left\{\left(\bigcap_{j=1}^{3}\begin{bmatrix} t^3 \\ r^2(\| \diamond (R_{j3}, R_{j4})) \end{bmatrix}\right) \bigcup \begin{bmatrix} t^3 \\ r^2(\| \diamond (R_{43}, R_{44})) \end{bmatrix}\right\} = 5$$

(3) Calculate DOF of the PM

According to the DOF formula (Eq. 6.8a in Chap. 6), DOF of the PM is

$$F = \sum_{i=1}^{m} f_i - \sum_{j=1}^{v} \xi_{L_j} = 20 - (6 + 5 + 5) = 4$$

So, DOF of the parallel mechanism in Fig. 10.3a meets the design requirement.

Step 4 **Select driving pairs**

Determine whether all driving pairs can be located on the same platform according to the criteria for driving pair selection (Sect. 6.3.2 in Chap. 6).

Example 10.4 Check whether the four R pairs on the base platform can be selected as driving pairs simultaneously for the parallel mechanism in Fig. 10.3a.

(1) Suppose the four R pairs $(R_{11}, R_{21}, R_{31}, R_{41})$ to be driving pairs and lock them, a new PM can be obtained. Topological structure of each new branch is

$$SOC\{-R_{j2} \| \widehat{R_{j3} \perp R_{j4}} \| R_{j5}-\}, \quad \mathrm{j} = 1, 2, 3, 4.$$

(2) According to Eq. (4.4) in Chap. 4, POC set of each new branch is

$$M_{bj} = \begin{bmatrix} t^3 \\ r^2(\| \diamond (R_{j3}, R_{j4}) \end{bmatrix}, \quad \mathrm{j} = 1, 2, 3, 4.$$

(3) Determine DOF of the new PM

Substitute POC set of each new branch into Eq. (6.8b) in Chap. 6 and obtain ξ_{L_j} as follows:

$$\xi_{L_1} = \dim\{M_{b_1} \bigcup M_{b_2}\}$$
$$= \dim\left\{\begin{bmatrix} t^3 \\ r^2(\|\diamondsuit(R_{13}, R_{14})) \end{bmatrix} \bigcup \begin{bmatrix} t^3 \\ r^2(\|\diamondsuit(R_{23}, R_{24})) \end{bmatrix}\right\} = 6$$

$$\xi_{L_2} = \dim\{(\bigcap_{j=1}^{2} M_{b_j}) \bigcup M_{b_3}\}$$
$$= \dim\left\{\bigcap_{j=1}^{2} \begin{bmatrix} t^3 \\ r^2(\|\diamondsuit(R_{j3}, R_{j4})) \end{bmatrix} \bigcup \begin{bmatrix} t^3 \\ r^2(\|\diamondsuit(R_{33}, R_{34})) \end{bmatrix}\right\} = 5$$

$$\xi_{L_3} = \dim\left\{(\bigcap_{j=1}^{3} M_{b_j}) \bigcup M_{b_4}\right\}$$
$$= \dim\left\{\left(\bigcap_{j=1}^{3} \begin{bmatrix} t^3 \\ r^2(\|\diamondsuit(R_{j3}, R_{j4})) \end{bmatrix}\right) \bigcup \begin{bmatrix} t^3 \\ r^2(\|\diamondsuit(R_{43}, R_{44})) \end{bmatrix}\right\} = 5$$

Then substitute ξ_{L_j} into Eq. (6.8a) in Chap. 6 and obtain DOF of the new PM

$$F^* = \sum_{i=1}^{m} f_i - \sum_{j=1}^{v} \xi_{L_j} = 16 - (6 + 5 + 5) = 0$$

(4) Since DOF of the obtained new PM is $F^* = 0$, the 4 R pairs on the base platform can be selected as driving pairs simultaneously according to the criteria for driving pair selection in Sect. 6.3.2 in Chap. 6.

For No. 1 branch combination scheme in Table 10.2, the branch assembling schemes have been determined (step 2), the DOFs have been checked (step 3) and the 4 R pairs can be selected as driving pairs simultaneously (step 4). So the (3T-1R) PMs satisfying the topological design requirement have been obtained.

Similarly, all other three parallel mechanisms in Fig. 10.3 can be checked as per the above procedure. The four PM structure types in Fig. 10.3 comply with the design requirement. Their basic parameters are listed in Table 10.3.

Table 10.3 Basic parameters of (3T-1R) PMs

No.	PM_S	Number of independent loops	Number of pairs	Number of links	Number of branches	Number of SOC branches	Number of HSOC branches
1	Figure 10.3a	$v = 3$	20	18	4	4	0
2	Figure 10.3b						
3	Figure 10.3c						
4	Figure 10.3d		16	14			
5	Figure 10.4a	$v = 7$	28	22		0	4
6	Figure 10.4b	$v = 5$	20	16		2	2
7	Figure 10.5	$v = 7$	16	10		2	2
8	Figure 10.6	$v = 4$	16	13	3	2	1
9	Figure 10.7a	$v = 7$	22	16	2	0	2
10	Figure 10.7b	$v = 5$	17	13	3	1	2
11	Figure 10.8	$v = 6$	19	14	2	1	1
12	Figure 10.9a	$v = 7$	32	26	4	0	4
13	Figure 10.9b	$v = 5$	22	18		2	2
14	Figure 10.10a	$v = 7$	28	22		0	4
15	Figure 10.10b	$v = 5$	20	16		2	2
16	Figure 10.11a	$v = 7$	32	26		0	4
17	Figure 10.11b	$v = 5$	22	18		2	2
18	Figure 10.12				2	0	2

10.2.5 *Structure Types of (3T-1R) Parallel Mechanism*

Follow the steps 1–4 in Sect. 10.2.4 for each branch combination scheme in Table 10.2, 18 types of (3T-1R) PMs containing no P pair are obtained, as shown in Figs. 10.3, 10.4, 10.5, 10.6, 10.7, 10.8, 10.9, 10.10, 10.11 and 10.12.

(1) For branch combination schemes 1 and 2 in Table 10.2, four (3T-1R) PM structure types are obtained, as shown in Fig. 10.3a–d. Their basic parameters are listed in Table 10.3.

(2) For branch combination schemes 3 and 4 in Table 10.2, two (3T-1R) PM structure types are obtained, as shown in Fig. 10.4a, b. Their basic parameters are listed in Table 10.3.

(3) For branch combination schemes 5 and 6 in Table 10.2, two (3T-1R) PM structure types are obtained, as shown in Figs. 10.5 and 10.6. Their basic parameters are listed in Table 10.3.

(4) For branch combination schemes 7 and 8 in Table 10.2, two (3T-1R) PM structure types are obtained, as shown in Fig. 10.7a [6–16] and Fig. 10.7b. Their basic parameters are listed in Table 10.3.

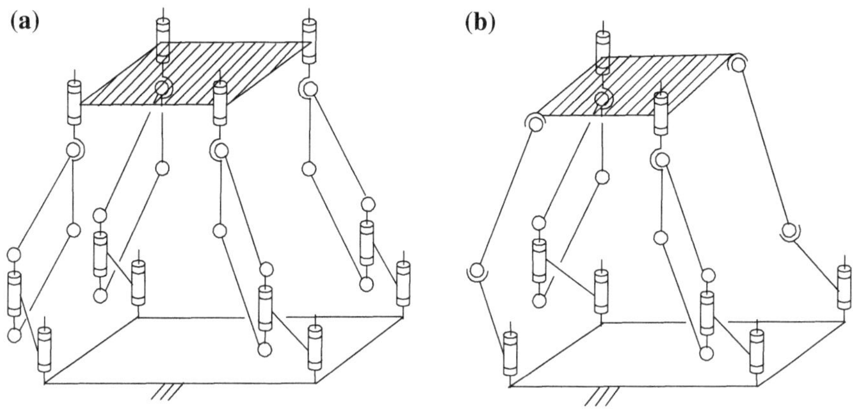

Fig. 10.4 (3T-1R) PMs corresponding to No. 3 and No. 4 branch assembling schemes

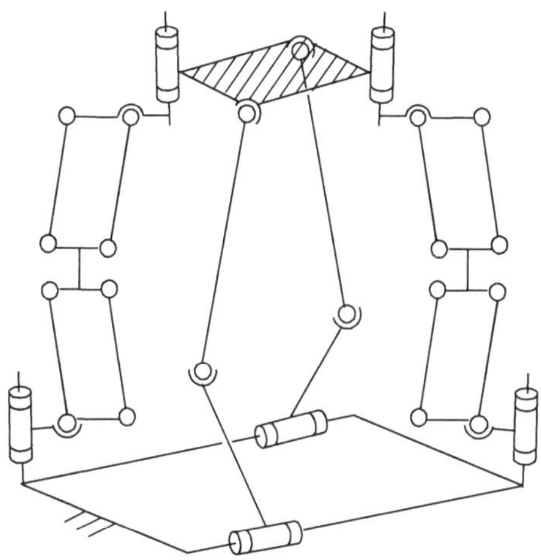

Fig. 10.5 (3T-1R) PM corresponding to No. 5 branch assembling scheme

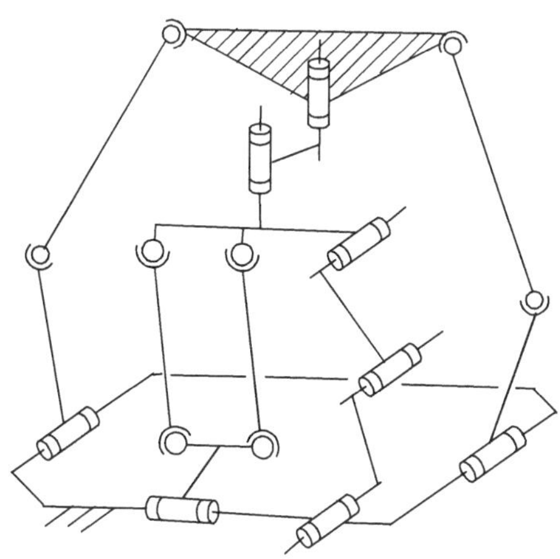

Fig. 10.6 (3T-1R) PM corresponding to No. 6 branch assembling scheme

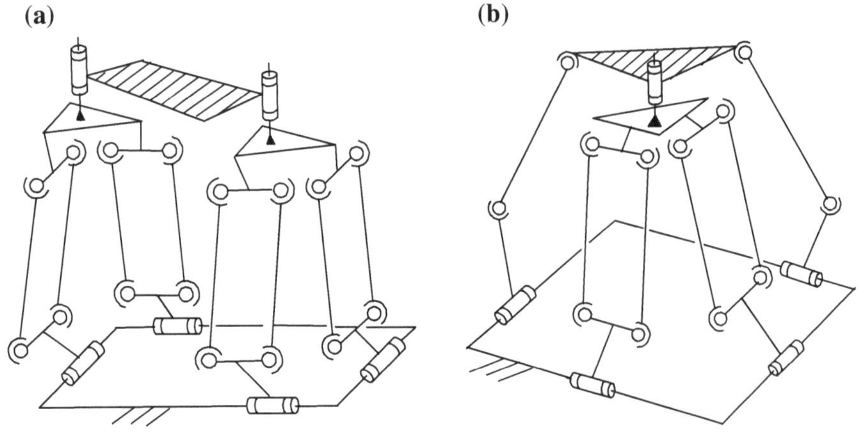

Fig. 10.7 (3T-1R) PMs corresponding to No. 7 and No. 8 branch assembling schemes

(5) For branch combination scheme 9 in Table 10.2, the obtained (3T-1R) PM is shown in Fig. 10.8 [4, 5]. Its basic parameters are listed in Table 10.3. Its HSOC branch contains a Delta mechanism.

(6) For branch combination schemes 10 and 11 in Table 10.2, two (3T-1R) PM structure types are obtained, as shown in Fig. 10.9a, b. Their basic parameters are listed in Table 10.3.

(7) For branch combination schemes 12 and 13 in Table 10.2, two (3T-1R) PM structure types are obtained, as shown in Fig. 10.10a, b. Their basic parameters are listed in Table 10.3.

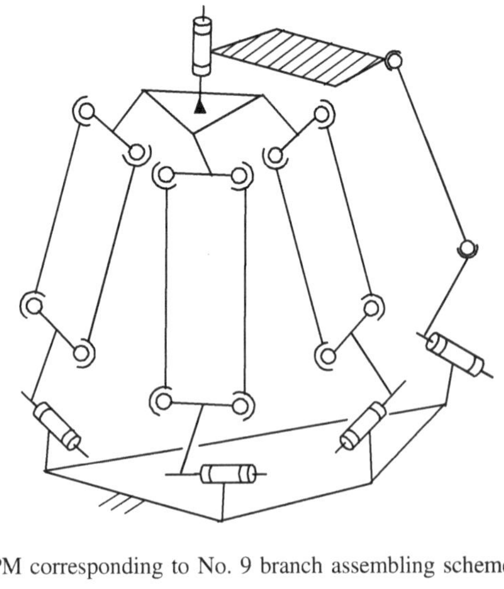

Fig. 10.8 (3T-1R) PM corresponding to No. 9 branch assembling scheme

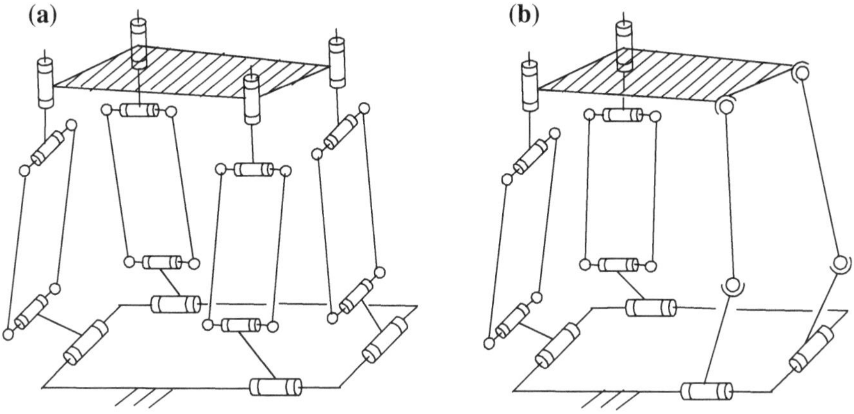

Fig. 10.9 (3T-1R) PMs corresponding to No. 10 and No. 11 branch assembling schemes

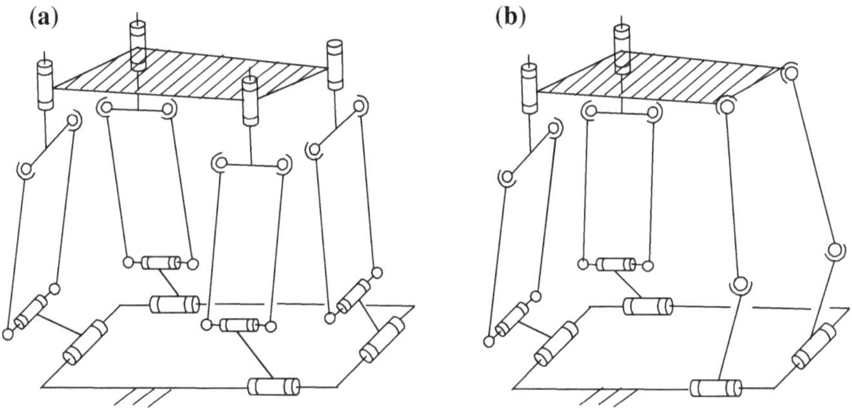

Fig. 10.10 (3T-1R) PMs corresponding to No. 12 and No. 13 branch assembling schemes

(8) For branch combination schemes 14 and 15 in Table 10.2, two (3T-1R) PM structure types are obtained, as shown in Fig. 10.11a [21] and Fig. 10.11b. Their basic parameters are listed in Table 10.3.

(9) For branch combination scheme 16 in Table 10.2, the obtained (3T-1R) PM is shown in Fig. 10.12. Its basic parameters are listed in Table 10.3.

For the above (3T-1R) PMs shown in Figs. 10.3, 10.4, 10.5, 10.6, 10.7, 10.8, 10.9, 10.10, 10.11 and 10.12, more PM structure types can be obtained through expansion in the following ways: (a) change the connection order among kinematic pairs (refer to Sect. 10.2.2), (b) combine several R and P pairs into C pair, U pair or S pair (refer to Sect. 10.2.2), (c) replace with topologically equivalent branches.

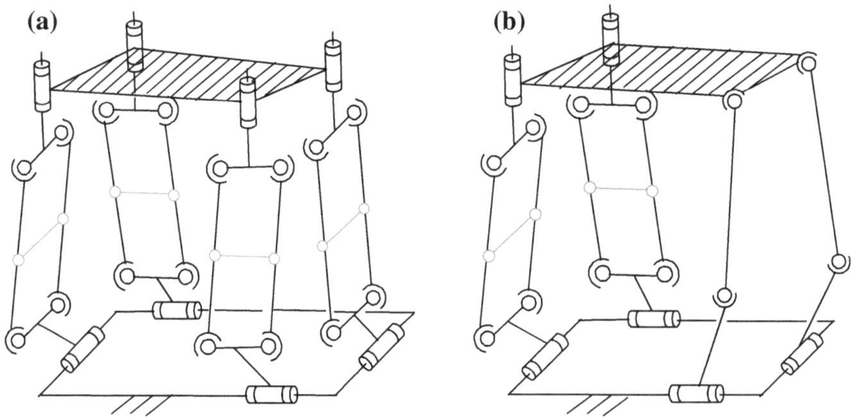

Fig. 10.11 (3T-1R) PMs corresponding to No. 14 and No. 15 branch assembling schemes

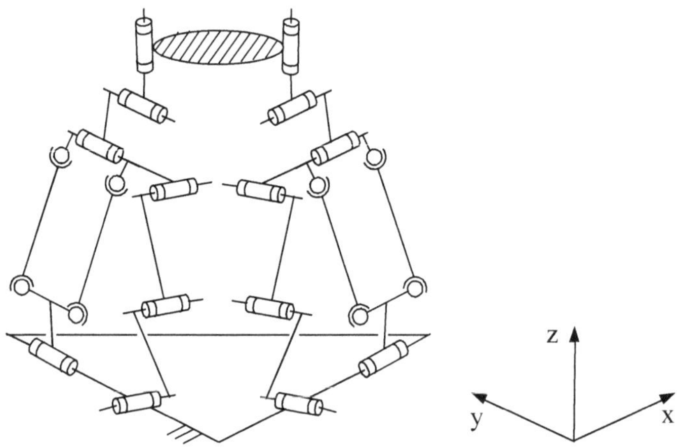

Fig. 10.12 (3T-1R) PM corresponding to No.16 branch assembling scheme

10.2.6 Performance Analysis of PM Structure Types Based on Topological Characteristics

As discussed in Sect. 9.6 in Chap. 9, topological characteristics of a parallel mechanism (Table 9.4 in Chap. 9) have close relations with its basic functions and kinematic (dynamic, control) performances.

Based on the topological characteristics listed in Table 9.4, the above 18 structure types of the (3T-1R) PM are analyzed. The result can be used for PM structure type assessment, classification and optimization.

Example 10.5 Determine topological characteristics of the parallel mechanism in Fig. 10.3a.

The topological characteristics can be determined as follows:

(1) According to Eq. (5.3) in Chap. 5, POC set of its movable platform is $M_{Pa} = \begin{bmatrix} t^3 \\ r^1 \end{bmatrix}$. The rotation axis is normal of the base platform (refer to step 1 in Sect. 10.2.4).

(2) Its DOF is DOF = 4 (refer to step 3 in Sect. 10.2.4).

(3) According to Eq. (3.10) in Chap. 3, dimension of its POC set is $\dim\{M_{Pa}\} = 4$.

(4) It contains no passive DOF.

(5) According to Eq. (6.10) in Chap. 6, its number of independent displacement equations is $\xi_{mech} = 6 + 5 + 5 = 16$ (refer to step 3 in Sect. 10.2.4).

(6) Number of over-constraint (Eq. (6.11) in Chap. 6): $N_{ov} = 18 - 16 = 2$.

(7) Redundancy (Eq. (6.12) in Chap. 6): $D_{re} = 0$.

(8) According to the inactive pair determination criteria (Sect. 6.3.1 in Chap. 6), it contains no inactive pair.

(9) According to the driving pair selection criteria, the four R pairs on the base platform can be selected as driving pairs simultaneously (refer to step 4 in Sect. 10.2.4).

(10) Determine the AKC it contained and corresponding coupling degree.

① Determine SOC_1

According to PM structure decomposition method (Sect. 7.4.1 in Chap. 7), the first SOC is

$$SOC_1\{-R_{11}\|R_{12}\|\widehat{R_{13}\perp R_{14}}\|R_{15}\|R_{25}\|\widehat{R_{24}\perp R_{23}}\|R_{22}\|R_{21}-\},$$
$$(R_{11} \text{ and } R_{21} \text{ are driving pairs})$$

② Determine the constraint degree of SOC_1

Since R_{11} and R_{21} are driving pairs and $\xi_{L_1} = 6$ (refer to step 3 in Sect. 10.2.4), there is (refer to Eq. (7.6) in Chap. 7)

$$\Delta_1 = \sum_{i=1}^{m_1} f_i - I_1 - \xi_{L_1} = 10 - 2 - 6 = +2$$

③ Determine SOC_2

According to PM structure decomposition method (Sect. 7.4.1 in Chap. 7), the second SOC is

$$SOC_2\{-R_{31}\|R_{32}\|\widehat{R_{33}\perp R_{34}}\|R_{35}-\}, (R_{31} \text{ is a driving pair})$$

④ Determine the constraint degree of SOC_2
Since R_{31} is a driving pair and $\xi_{L_2} = 5$ (refer to step 3 in Sect. 10.2.4), there is (refer to Eq. (7.6) in Chap. 7)

$$\Delta_2 = \sum_{i=1}^{m_2} f_i - I_2 - \xi_{L_2} = 5 - 1 - 5 = -1$$

⑤ Determine SOC_3
According to PM structure decomposition method (Sect. 7.4.1 in Chap. 7), the third SOC is

$$SOC_3\{-R_{41}\|R_{42}\|\overbrace{R_{43}\perp R_{44}}\|R_{45}-\}, (\mathrm{R}_{31} \text{ is a driving pair})$$

⑥ Determine the constraint degree of SOC_3
Since R_{41} is a driving pair and $\xi_{L_3} = 5$ (refer to step 3 in Sect. 10.2.4), there is (refer to Eq. (7.6) in Chap. 7)

$$\Delta_3 = \sum_{i=1}^{m_2} f_i - I_2 - \xi_{L_2} = 5 - 1 - 5 = -1$$

⑦ Determine AKC and coupling degree κ
According to AKC determination method (Sect. 7.4.2 and Eq. (7.9) in Chap. 7), this PM contains only one $AKC[v = 3, \kappa = 2]$. Its coupling degree is

$$k = \frac{1}{2}\min\{\sum_{j=1}^{v} |\Delta_j|\} = \frac{1}{2}(|+2| + |-1| + |-1|) = 2$$

(11) Type of DOF

Since this PM contains only one AKC, it has full-cycle DOF according to DOF type determination criteria (Sect. 7.5 in Chap. 7).

(12) Motion input-output decoupling

Since this PM has full-cycle DOF, it is not motion decoupled according to the motion decoupling principle (Sect. 7.6 in Chap. 7).

(13) This PM has 4 branches with identical structures.

The above topological characteristics of the PM in Fig. 10.3a are listed in Table 10.4. Topological characteristics of other 17 types of PMs can be determined following a similar procedure, as shown in Table 10.4.

Table 10.4 Topological characteristics of (3T-1R) PMs

No.	PM_S	M_{Pa}, Dim $\{M_{Pa}\}$	DOF	Local DOF	Inactive pair	ξ_{mech}	N_{ov}	D_{re}	Driving pair allocation	$AKC[v, \kappa]$	DOF type	Motion decoupling	Branch structure
1	Figure 10.3a	$M_{Pa} = \begin{bmatrix} t^3 \\ r^1 \end{bmatrix}$ Dim $\{M_{Pa}\} = 4$	4	0	0	16	2	0	Base platform	$AKC[v = 3, \kappa = 2]$	Full-cycle DOF	No decoupling	Same
2	Figure 10.3b												
3	Figure 10.3c												
4	Figure 10.3d			2		18	0						2 types
5	Figure 10.4a			0		24	18						Same
6	Figure 10.4b			2		22	8						2 types
7	Figure 10.5					28	14						
8	Figure 10.6					23	1			$AKC_1[v = 1, \kappa = 0]$ $AKC_2[v = 2, \kappa = 1]$	Partial DOF	Partial decoupling	
9	Figure 10.7a			8		38	4			$AKC[v = 3, \kappa = 2]$	Full-cycle DOF	No decoupling	Same
10	Figure 10.7b			6		30	1			$AKC[v = 3, \kappa = 1]$			2 types
11	Figure 10.8			7		34	2			$AKC_1[v = 2, \kappa = 1]$ $AKC_2[v = 1, \kappa = 0]$	Partial DOF	Partial decoupling	2 types
12	Figure 10.9a			0		28	14			$AKC[v = 3, \kappa = 2]$	Full-cycle DOF	No decoupling	Same
13	Figure 10.9b			2		24	6						2 types
14	Figure 10.10a			0		40	2						Same
15	Figure 10.10b			2		30	0						2 types
16	Figure 10.11a			0		40	2						Same
17	Figure 10.11b			2		30	0						2 types
18	Figure 10.12			2		28	2			$3 - AKC[v = 1, \kappa = 0]$	Partial DOF	Partial decoupling	Same

Note ξ_{mech}—Number of independent displacement equations, N_{ov}—number of overconstraints, D_{re}—redundant DOF

10.2.7 Classification of (3T-1R) PM Structure Types Based on Performance Analysis

The (3T-1R) PMs are classified according to their topological characteristics in Table 10.4, as shown in Table 10.5. It will help the designers select the one which best fits their application.

Type-1 PMs formed by SOC branches

As shown in Table 10.5, the PMs in Fig. 10.3a–d are this type of PMs.

The main characteristics of this type of PM include:

(1) The branch structure is relatively simple. It may have a relatively larger workspace. But the rigidity is relatively lower than that of type-2 and type-3 PMs.

(2) It contains only one $AKC[v = 3, \kappa = 2]$. Its forward displacement analysis is relatively complex (refer to Chap. 7).

(3) The PMs shown in Fig. 10.3a–c are overconstrained mechanisms. The PM in Fig. 10.3d is a non-overconstrained mechanism.

(4) It is not motion-decoupled.

Type-2 PMs formed by HSOC branches

As shown in Table 10.5, the PMs in Figs. 10.4a, 10.7a, 10.9a, 10.10a, 10.11a and 10.12 are this type of PMs.

The main characteristics of this type of PMs include:

(1) The branch structure is complicated. So the PM has high rigidity. But the interference among branches may reduce the PM's workspace, especially the PMs formed by only two HSOCs (Figs. 10.7a and 10.12). So, an additional rotation mechanism is needed in order to expand the moving platform's rotation range [11–20].

(2) Most PMs of this type contain only one $AKC[v = 3, \kappa = 2]$. Its forward displacement analysis is relatively complex. But the PM shown in Fig. 10.12 contains three $AKCs[v = 3, \kappa = 0]$ and closed-form solutions can be obtained for its forward kinematics.

(3) It is an overconstrained mechanism.

(4) Most PMs of this type are not motion-decoupled. But the PM shown in Fig. 10.12 is partially decoupled (refer to Type-4 for detail).

Table 10.5 Four types of (3T-1R) PMs

Type-1	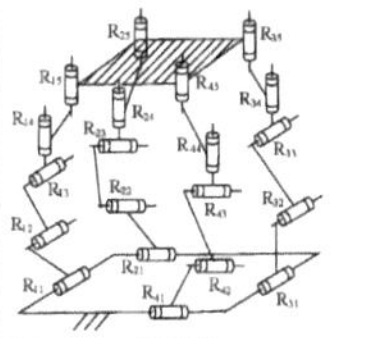 Figure 10.3a	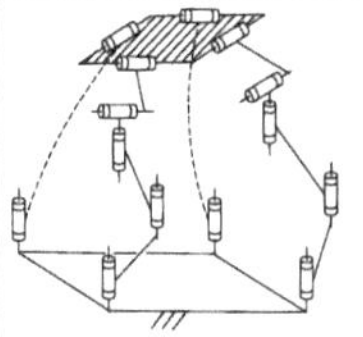 Figure 10.3b	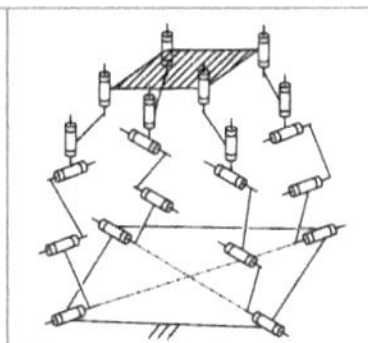 Figure 10.3c	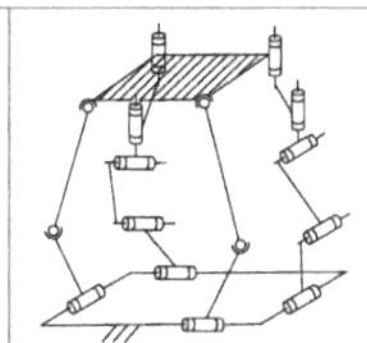 Figure 10.3d
Type-2	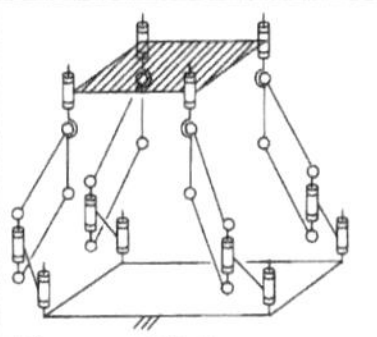 Figure 10.4a	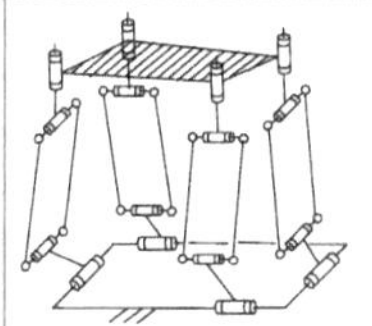 Figure 10.9a	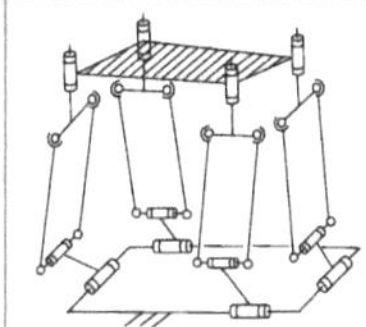 Figure 10.10a	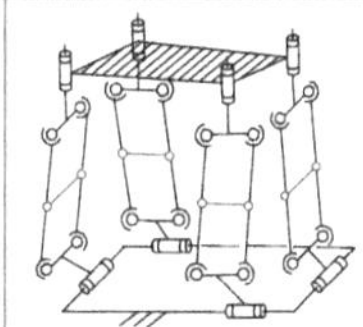 Figure 10.11a [21]
	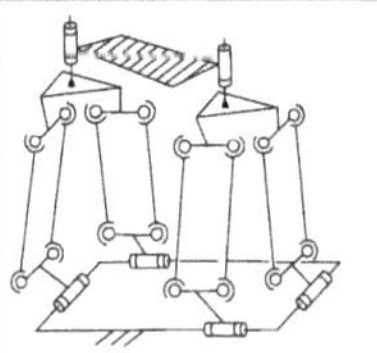 Figure 10.7a [6, 7]	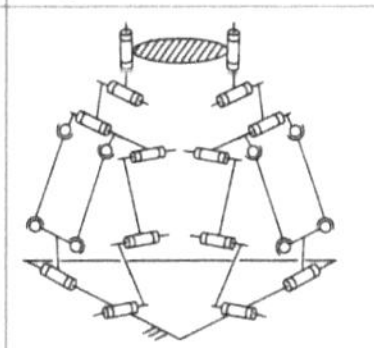 Figure 10.12	Note: Figure 10.7a—H4 PM [6, 7] Figure 10.8—ABB PM [4, 5] Figure 10.11a—X4 PM [21]	
Type-3	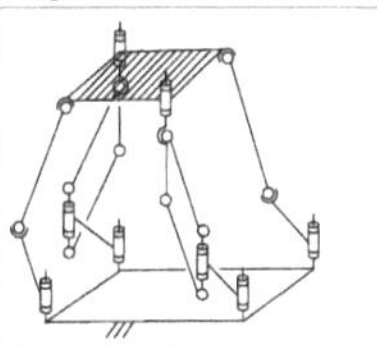 Figure 10.4b	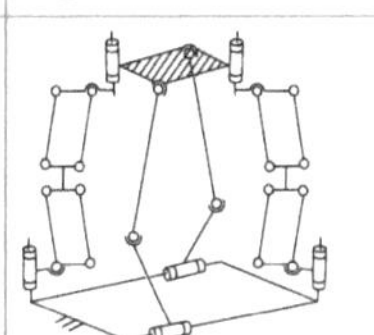 Figure 10.5	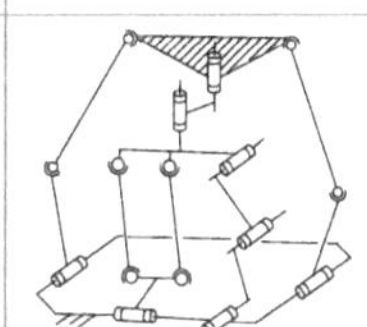 Figure 10.6	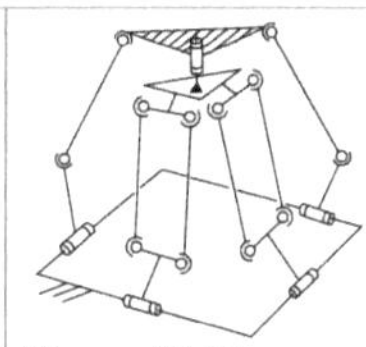 Figure 10.7b
	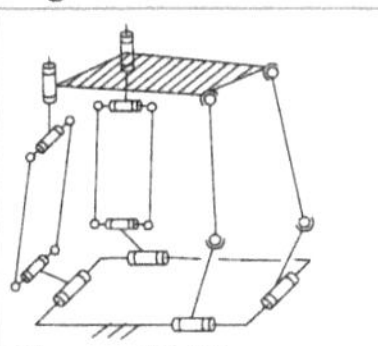 Figure 10.9b	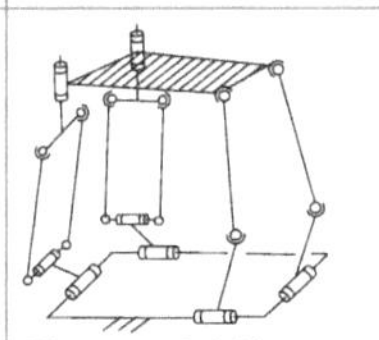 Figure 10.10b	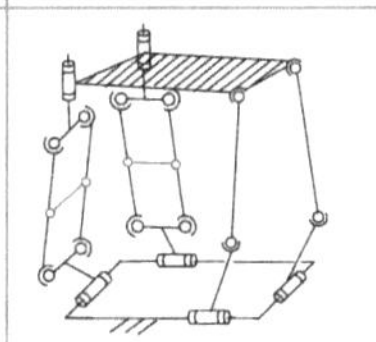 Figure 10.11b	
Type-4	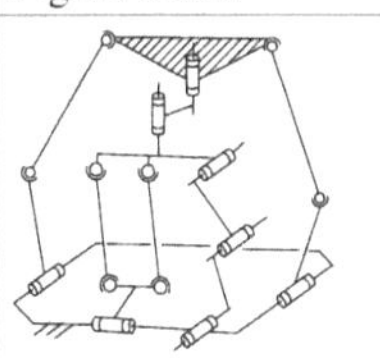 Figure 10.6	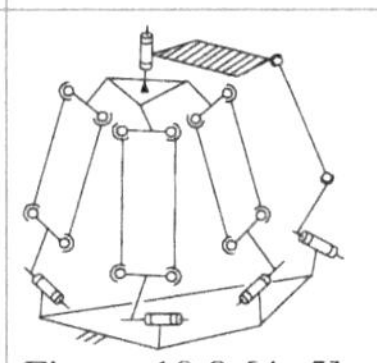 Figure 10.8 [4, 5]	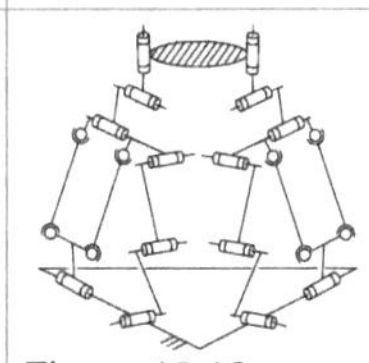 Figure 10.12	

Type-3 PMs with HSOC branches and SOC branches

As shown in Table 10.5, the PMs in Figs. 10.4b, 10.5, 10.6, 10.7b, 10.9b, 10.10b and 10.11b are this type of PMs.

The main characteristics of this type of PM include:

(1) Since it has 2 simple SOC branches, it may have a relatively larger workspace when compared with type-2 PMs. In addition, it has a relatively higher rigidity when compared with type-1 PMs since it contains HSOC branches.

(2) Most PMs of this type contain only one $AKC[v = 3, \kappa = 2]$, so its forward displacement analysis is relatively complex. But, the PMs in Figs. 10.6 and 10.7b contain $AKC[v = 3, \kappa = 1]$, so its forward displacement analysis is relatively simple.

(3) Most PMs of this type are overconstrained mechanisms. But the PMs shown in Figs. 10.10b and 10.11b are non-overconstrained mechanisms.

(4) Most PMs of this type are not motion-decoupled. But, the PM in Fig. 10.6 is partially decoupled (refer to Type-4 for detail).

Type-4 Partially decoupled PMs.

As shown in Table 10.5, the PMs in Figs. 10.6, 10.8 and 10.12 are this type of PMs.

(1) The main characteristics of the PM in Fig. 10.6 include:

  (a) The output Z-axis (normal line of the fixed platform) translation of the moving platform is related only to the two inputs in the HSOC branch and has no relation with the input in the SOC branch. So, this PM is partially decoupled.

  (b) Two simple SOC branches help to expand the PM's workspace.

  (c) This PM contains two AKCs: $AKC_1[v = 1, \kappa = 0]$ and $AKC_2[v = 2, \kappa = 1]$. Its forward displacement analysis is relatively simple.

  (d) It is an overconstrained mechanism.

(2) The main characteristics of the PM in Fig. 10.8 include:

  (a) The PM has a moving platform (manipulator) and a sub-PM (Delta mechanism). The 3 translations of the manipulator depend only on the three inputs of the sub-PM (Delta mechanism. So, this PM is partially decoupled.

  (b) A simple SOC branch helps to expand the PM's workspace.

  (c) This PM contains 2 AKCs: $AKC_1[v = 2, \kappa = 1]$ and $AKC_2[v = 1, \kappa = 0]$. Its forward displacement analysis is relatively simple.

  (d) It is an overconstrained mechanism.

(3) The main characteristics of the PM in Fig. 10.12 include:

(a) The output translation in Y direction is related only to the two inputs in the left-side HSOC. The output translation in X direction is related only to the two inputs in the right-side HSOC. The translation in Z direction is related to all the four inputs. So, this PM is partially decoupled.

(b) The interference between two complex HSOCs may reduce the PM's workspace.

(c) This PM contains 3 AKCs: $AKC[v = 1, \kappa = 0]$. Its forward displacement analysis is very easy.

(d) It is an overconstrained mechanism.

Among the 18 different (3T-1R) PMs obtained, two PMs (Figs. 10.7a [6, 7] and 10.8 [4, 5]) have already been used in SCARA parallel robots. Application of the PM shown in Fig. 10.11a is under experiment [21]. For the other 15 PMs, kinematic and dynamic analyses may be conducted in order to find any new PMs with good performance which can be used in SCARA parallel robots.

According to above PM classification, a type-3 PM has a higher rigidity than a type-1 PM and a larger workspace than a type-2 PM. So, special attentions shall be paid to type-3 PMs, especially the seven PMs shown in Figs. 10.3d, 10.4b, 10.5, 10.6, 10.7b, 10.9b and 10.11b. The inverse position analysis, workspace, singular position, forward position analysis of the PM shown in Fig. 10.7b have already been systematically analyzed [43].

10.3 Summary

(1) The procedure for topology design of (3R-1R) PM is discussed in detail based on the systematic method for topology design of parallel mechanisms introduced in Chap. 9. The design process includes two stages. The first stage involves traditional structure synthesis which results in 18 structure types of (3R-1R) PMs which contain no P pairs. The second stage includes performance analysis and classification of the obtained PM structure types which may help designer to select the most suitable structure type.

(2) Key steps of topology design for (3R-1R) PMs include:

(a) Determine structure types of the SOC branch containing only R and P pairs according to the method for structure synthesis of SOC branch described in Chap. 8. Then obtain corresponding HSOC branches (Sect. 10.2.2) through topologically equivalent principle (Sect. 9.3.3 in Chap. 9).

(b) Determine the geometrical condition for assembling branches between two platforms according to corresponding basic formulas (Eqs. (9.2a)–(9.5f) in

Chap. 9). Then determine different assembling schemes for the same geometrical condition (Sect. 10.2.4).

(c) Performance analysis and classification of the obtained PM structures are conducted based on topological characteristics of PMs (Table 9.4 in Chap. 9). The result can be used for PM structure assessment and optimization (Sects. 10.2.6 and 10.2.7).

(3) The obtained (3R-1R) PMs are classified into four types:

(a) PMs containing only SOC branches, such as the PMs shown in Fig. 10.3a–d).

(b) PMs containing only HSOC branches, such as the PMs shown in Figs. 10.4a, 10.7a, 10.9a, 10.10a, 10.11a and 10.12.

(c) PMs containing both HSOC and SOC branches, such as the PMs shown in Figs. 10.4b, 10.5, 10.6, 10.7b, 10.9b, 10.10b and 10.11b.

(d) PMs with partial motion decoupling property, such as the PMs shown in Figs. 10.6, 10.8 and 10.12.

(4) The type-3 PM has a higher rigidity than a type-1 PM and a larger workspace than a type-2 PM. So, special attentions shall be paid to type-3 PMs.

References

1. Schoenflies AM (1893) La géométrie du mouvement. Gauthier-Villars, Paris
2. Mitsubishi SCARA robots-RH Series [EB/OL] (1981) http://www.meau.com/eprise/main/sites/public/Products/Robots/Robot_Videos/default
3. Clavel R (1990) Device for the movement and positioning of an element in space. US Patent 4 976 582
4. ABB. IRB 360 FlexPicker [EB/OL]. http://new.abb.com/products/robotics/industrial-robots/irb-360
5. Fanuc. M-1iA series robots [EB/OL]. http://robot.fanucamerica.com/products/robots/product.aspx
6. Pierrot F, Company O (1999) H4: a new family of 4-DOF parallel robots. In: IEEE/ASME international conference on advanced intelligent mechatronics, Atlanta, GA, pp 508–513, 19–23 Sept 1999
7. Company O, Marquet F, Pierrot F (2003) A new high-speed 4-DOF parallel robot synthesis and modeling issues. IEEE Trans Robot Autom 19(3):411–420
8. Krut S, Company O, Benoit M et al (2003) I4: a new parallel mechanism for SCARA motions issues. In: Proceedings of the IEEE international conference on robotics and automation. IEEE, Taiwan, pp 1875–1880, 14–19 Sept 2003
9. Nabat V, de la O Rodriguez M, Company O, Krut S, Pierrot F (2005) Par4: very high speed parallel robot for pick-and-place. In: IEEE/RSJ international conference on intelligent robots and systems (IROS 2005), Edmonton, Canada, pp 553–558

10. Angeles J, Caro S, Khan W, Morozov A (2006) Kinetostatic design of an innovative Schönflies-motion generator. Proc Inst Mech Eng Part C J Mech Eng Sci 200(7):935–943
11. Pierrot F, Nabat V, Company O et al (2009) Optimal design of a 4-DOF parallel manipulator: from academia to industry. IEEE Trans Rob 25(2):213–224
12. Adept Quattro s650H [EB/OL]. http://www.adept.com/products/robots/parallel/quattro-s650h/general
13. Altuzarra O, Hernandez A, Salgado O, Angeles J (2009) Multiobjective optimum design of a symmetric parallel Schonflies-motion generator. ASME J Mech Des 131(3):031002
14. Pierrot F, Nabat V, Company O, Krut S, Poignet P (2009) Optimal design of a 4-DOF parallel manipulator: from academia to industry. IEEE Trans Rob 25(2):213–224
15. Kim SM, Kim W, Yi BJ (2009) Kinematic analysis and optimal design of a 3T1R type parallel mechanism. In: IEEE international conference on robotics and automation (ICRA'09), Kobe, Japan, pp 2199–2204, 12–17 May 2009
16. Ancuta A, Company O, Pierrot F (2010) Design of Lambda-Quadriglide: a new 4-DOF parallel kinematics mechanism for Schonflies motion. ASME Paper No. DETC2010–28715
17. Briot S, Bonev IA (2010) Pantopteron-4: a new 3T1R decoupled parallel manipulator for pick-and-place applications. Mech Mach Theory 45(5):707–721
18. Corbel D, Gouttefarde M, Company O, Pierrot F (2010) Actuation redundancy as a way to improve the acceleration capabilities of 3T and 3T1R pick-and-place parallel manipulators. ASME J Mech Rob 2(4):041002
19. Altuzarra O, Sandru B, Pinto C, Petuya V (2011) A symmetric parallel schonflies-motion manipulator for pick-and-place operations. Robotica 29(6):853–862
20. Liu ST, Huang T, Mei JP, Zhao XM, Wang PF, Chetwynd DG (2012) Optimal design of a 4-DOF scara type parallel robot using dynamic performance indices and angular constraints. ASME J Mech Rob 4(3):0310053
21. Xie F, Liu X-J (2015) Design and development of a high-speed and high-rotation robot with four identical arms and a single platform. ASME JMR 7:041015-1–041015-12
22. Hervé JM (1995) Design of parallel manipulators via the displacement group. In: Proceedings of the 9th world congress on the theory of machines and mechanisms, Milan, Italy, pp 2079–2082
23. Li QC, Herve JM (2009) Parallel mechanisms with bifurcation of Schonflies motion. IEEE Trans Rob 25(1):158–164
24. Salgado O, Altuzarra O, Petuya V, Hernandez A (2008) Synthesis and design of a novel 3T1R fully-parallel manipulator. ASME J Mech Des 130:042305-1–042305-8
25. Li Z, Lou Y, Zhang Y, Liao B, Li Z (2013) Type synthesis, kinematic analysis, and optimal design of a novel class of Schonflies-motion parallel manipulators. IEEE Trans Autom Sci Eng 10(3):674–686
26. Huang Z, Li QC (2002) General methodology for type synthesis of symmetrical lower-mobility parallel manipulators and several novel manipulators. Int J Robot Res 21 (2):131–145
27. Kong X, Gosselin CM (2004) Type Synthesis of 3T1R 4-DOF Parallel Manipulators Based on Screw Theory. IEEE Trans. Rob. Autom. 20(2):181–190
28. Kong X, Gosselin CM (2007) Type synthesis of parallel mechanisms, vol 33. Springer tracts in advanced robotics
29. Amine S, Masouleh MT, Caro S, Wenger P, Gosselin C (2012) Singularity conditions of 3T1R parallel manipulators with identical limb structures. ASME J Mech Rob 4(1):011011
30. Gogu G (2007) Structural synthesis of fully-isotropic parallel robots with Schönflies motions via theory of linear transformations and evolutionary morphology. Eur J Mech A/Solids 26:242–269
31. Gogu G (2008) Structural synthesis of parallel robots: Part 1: Methodology. Springer, Dordrecht
32. Yang T-L, Jin Q, Liu A-X, Yao F-H, Luo Y-F (2001) Structure synthesis of 4-DOF (3-Translation and 1-Rotation) parallel robot mechanisms based on the units of

single-opened-chain. In: Proceedings of the ASME design engineering technical conference, DETC2001/DAC-21152
33. Jin Q, Yang T-L (2002) Structure synthesis of parallel manipulators with two-dimensional translation and one-dimensional rotation. In: Proceedings of ASME design engineering technical conference, Montreal, MECH-34307
34. Yang T-L, Jin Q, Liu A-X, Shen H-P, Luo Y-F (2002) Structural synthesis and classification of the 3-DOF translation parallel robot mechanisms based on the unites of single-open chain. Chin J Mech Eng 38(8):31–36. doi:10.3901/JME.2002.08.031
35. Jin Q, Yang T-L (2004) Theory for topology synthesis of parallel manipulators and its application to three-dimension-translation parallel manipulators. ASME J Mech Des 126:625–639
36. Yang T-L (2004) Theory of topological structure for robot mechanisms. China Machine Press, Beijing
37. Meng X, Gao F, Wu S, Ge J (2014) Type synthesis of parallel robotic mechanisms: framework and brief review. Mech Mach Theory 78:177–186
38. Yang T-L, Liu A-X, Luo Y-F et al (2009) Position and orientation characteristic equation for topological design of robot mechanisms. ASME J Mech Des 131:021001-1–21001-17
39. Yang T-L, Liu A-X, Shen H-P et al (2013) On the correctness and strictness of the POC equation for topological structure design of robot mechanisms. ASME J. Mech Robot 5 (2):021009-1–021009-18
40. Yang T-L, Sun D-J (2012) A general DOF formula for parallel mechanisms and multi-loop spatial mechanisms. ASME J Mech Robot 4(1):011001-1–011001-17
41. Yang T-L, Liu A-X, Shen H-P et al (2015) Composition principle based on SOC unit and coupling degree of AKC for general spatial mechanisms. The 14th IFToMM world congress, Taipei, Taiwan, 25–30 Oct 2015. doi:10.6567/IFToMM.14TH.WC.OS13.135
42. Yang T-L, Liu A-X, Shen H-P et al (2012) Theory and application of robot mechanism topology. China Science Press, Beijing
43. Yang T-L, Liu A-X, Shen H-P, Hang L-B (2016) Topological structure synthesis of 3T1R parallel mechanism based on POC equations. In: Proceedings of 9th international conference on intelligent robotics and applications, 147–161. doi:10.1007/978-3-319-43506-0_13
44. Shen H-P, Shao G-W, Deng J-M, Yang T-L (2017) A novel 3T1R parallel robot 2PaRSS: design and kinematics. In: Proceedings of ASME 2017 international design engineering technical conferences, DETC2017-67265
45. Yang T-L, Liu A-X, Shen H-P, Hang L-B (2017) Topological structural synthesis of 3T-1R parallel mechanism based on POC equations. Chin J Mech Eng 53(21):54–64. doi:10.3901/JME.2017.21.054

MIX
Papier aus verantwortungsvollen Quellen
Paper from responsible sources
FSC® C105338

If you have any concerns about our products,
you can contact us on
ProductSafety@springernature.com

In case Publisher is established outside the EU,
the EU authorized representative is:
Springer Nature Customer Service Center GmbH
Europaplatz 3, 69115 Heidelberg, Germany

Printed by Libri Plureos GmbH
in Hamburg, Germany